Green Nanotechnology: The Latest Innovations, Knowledge Gaps, and Future Perspectives

Green Nanotechnology: The Latest Innovations, Knowledge Gaps, and Future Perspectives

Editors

Roberto Martins
Olga Barbara Kaczerewska

MDPI • Basel • Beijing • Wuhan • Barcelona • Belgrade • Manchester • Tokyo • Cluj • Tianjin

Editors
Roberto Martins
CESAM—Centre for
Environmental and Marine
Studies and Department of
Biology
University of Aveiro
Aveiro
Portugal

Olga Barbara Kaczerewska
CICECO—Aveiro Institute of
Materials and Department of
Materials and Ceramic
Engineering
University of Aveiro
Aveiro
Portugal

Editorial Office
MDPI
St. Alban-Anlage 66
4052 Basel, Switzerland

This is a reprint of articles from the Special Issue published online in the open access journal *Applied Sciences* (ISSN 2076-3417) (available at: www.mdpi.com/journal/applsci/special_issues/ Innovative_Green_Nanotechnology).

For citation purposes, cite each article independently as indicated on the article page online and as indicated below:

LastName, A.A.; LastName, B.B.; LastName, C.C. Article Title. *Journal Name* **Year**, *Volume Number*, Page Range.

ISBN 978-3-0365-2009-4 (Hbk)
ISBN 978-3-0365-2008-7 (PDF)

© 2022 by the authors. Articles in this book are Open Access and distributed under the Creative Commons Attribution (CC BY) license, which allows users to download, copy and build upon published articles, as long as the author and publisher are properly credited, which ensures maximum dissemination and a wider impact of our publications.

The book as a whole is distributed by MDPI under the terms and conditions of the Creative Commons license CC BY-NC-ND.

Contents

About the Editors vii

Preface to "Green Nanotechnology: The Latest Innovations, Knowledge Gaps, and Future Perspectives" ix

Roberto Martins and Olga Barbara Kaczerewska
Green Nanotechnology: The Latest Innovations, Knowledge Gaps, and Future Perspectives
Reprinted from: *Appl. Sci.* **2021**, *11*, 4513, doi:10.3390/app11104513 1

Juliana G. Galvão, Raquel L. Santos, Ana Amélia M. Lira, Renata Kaminski, Victor H. Sarmento and Patricia Severino et al.
Stearic Acid, Beeswax and Carnauba Wax as Green Raw Materials for the Loading of Carvacrol into Nanostructured Lipid Carriers
Reprinted from: *Appl. Sci.* **2020**, *10*, 6267, doi:10.3390/app10186267 5

Marina M. Mennucci, Rodrigo Montes, Alexandre C. Bastos, Alcino Monteiro, Pedro Oliveira and João Tedim et al.
Nanostructured Black Nickel Coating as Replacement for Black Cr(VI) Finish
Reprinted from: *Appl. Sci.* **2021**, *11*, 3924, doi:10.3390/app11093924 19

Bogumił Brycki, Adrianna Szulc and Mariia Babkova
Synthesis of Silver Nanoparticles with Gemini Surfactants as Efficient Capping and Stabilizing Agents
Reprinted from: *Appl. Sci.* **2020**, *11*, 154, doi:10.3390/app11010154 33

Olga Kaczerewska, Isabel Sousa, Roberto Martins, Joana Figueiredo, Susana Loureiro and João Tedim
Gemini Surfactant as a Template Agent for the Synthesis of More Eco-Friendly Silica Nanocapsules
Reprinted from: *Appl. Sci.* **2020**, *10*, 8085, doi:10.3390/app10228085 47

Juliana Vitoria Nicolau dos Santos, Roberto Martins, Mayana Karoline Fontes, Bruno Galvão de Campos, Mariana Bruni Marques do Prado e Silva and Frederico Maia et al.
Can Encapsulation of the Biocide DCOIT Affect the Anti-Fouling Efficacy and Toxicity on Tropical Bivalves?
Reprinted from: *Appl. Sci.* **2020**, *10*, 8579, doi:10.3390/app10238579 61

About the Editors

Roberto Martins

Dr. Roberto Martins (Ph.D. in Biology, 2013) is an Assistant Researcher from CESAM & Department of Biology, University of Aveiro (Portugal), a member of the General Council of the University of Aveiro since July 2021, and he has been a Visiting Professor at the State University of São Paulo (UNESP, Brazil) and the University of São Paulo (USP, Brazil). Over the last 5 years, his main research interests have been related to assessments of the efficacy and ecotoxicity of innovative nanomaterials to control key societal problems (e.g., corrosion and fouling) and the development of nano-specific guidelines to fulfil regulatory needs. He is the (co-)author of 41 ISI papers (h-index=15), 1 book, and 1 book chapter. In the last 5 years, Dr. Martins has supervised 3 postdocs, and more than 20 Ph.D., MSc, and BSc students. He has been involved in several R&D projects, mostly related to (nano)ecotoxicology and outreach activities aimed at promoting public awareness.

Olga Barbara Kaczerewska

Dr. Olga Kaczerewska (Ph.D. in Chemistry, 2017) is a Researcher at Reckitt (Germany). Previously, she was a Researcher (2018–2021) at the University of Aveiro (Portugal). Over the last 5 years, her main research interests have been related to the development of "smart" maritime anticorrosion coatings, innovative anti-corrosion nanocontainers, and novel eco-friendly cationic gemini surfactants as efficient corrosion inhibitors. She is the (co-)author of 1 patent, 10 ISI papers (h-index=7), 1 book, and 2 book chapters.

Preface to "Green Nanotechnology: The Latest Innovations, Knowledge Gaps, and Future Perspectives"

Nanotechnology is a key enabling technology, bringing together chemists, biologists, physicists, and materials science engineers, among others. Due to its vast range of applications, uses have been proposed for addressing societal challenges. Not surprisingly, the use of nanostructured materials has raised health and environmental safety concerns, favoring the expansion of a sub-field dedicated to green and safe-by-design solutions. Novel solutions should minimize the environmental and human health risks of nanomaterials during their lifetime, e.g., through the replacement of toxic products or current processes by suitable eco-friendly alternatives. Green nanotechnology relies on the principles of green chemistry towards the sustainable design, manufacture, use, and end-of-life of nanomaterials.

Roberto Martins and Olga Barbara Kaczerewska
Editors

Green Nanotechnology: The Latest Innovations, Knowledge Gaps, and Future Perspectives

Roberto Martins [1,*] and Olga Barbara Kaczerewska [1,2]

1. CESAM—Centre for Environmental and Marine Studies and Department of Biology, University of Aveiro, 3810-193 Aveiro, Portugal; olga.kaczerewska@ua.pt
2. CICECO—Aveiro Institute of Materials and Department of Materials and Ceramic Engineering, University of Aveiro, 3810-193 Aveiro, Portugal
* Correspondence: roberto@ua.pt

Keywords: green nanotechnology; green chemistry; safe-by-design; sustainability; nanomaterials; nanoparticles; nanoecotoxicology; environmental risk assessment

1. Introduction

Nanotechnology is a key enabling technology bringing together chemists, biologists, physicists, and materials science engineers, among others [1–6]. It has been proposed for addressing societal challenges due to its vast range of applications, such as on nanomedicine, food, nanoelectronics, energy, packaging, composite materials, coatings, construction, agriculture, water treatment and environmental remediation [5,7,8]. Not surprisingly, the use of nanostructured materials has been raising health and environmental safety concerns [5,9,10], favoring the expansion of a sub-field dedicated to green and safe-by-design solutions [1,2,6]. Novel solutions should minimize environmental and human health risks of nanomaterials during their lifetime, e.g., through the replacement of toxic products or current processes by suitable eco-friendly alternatives [2,11]. Green nanotechnology relies on the principles of green chemistry towards a sustainable design, manufacture, use, and end-of-life of nanomaterials [11,12].

2. Latest Innovations and Insight on the Domain of Green Nanotechnology

This Special Issue, spread through five original research articles [2–6], aggregates innovative applications, products, technologies and processes beyond the state-of-the-art in several scientific green nanotechnology-related fields (e.g., drug delivery systems, antifouling nanoadditives and coatings for optical applications) as well as identifying some knowledge gaps on this domain. Research on sustainable production of nanomaterials based on safe-by-design approaches and (eco)toxicological assessment of novel nanomaterials was also provided (Figure 1).

Gemini surfactants are being proposed as promising eco-friendly replacements of state-of-the-art surfactants, for instance, to synthesize greener nanomaterials. In this domain, Brycki et al. [3] proposed an ecofriendly synthesis of AgNPs stabilized by gemini surfactants produced with a solvent-free method. The smallest AgNPs were obtained using the surfactant 16-6-16 as a stabilizing agent, molar ratio nAg:nGemini = 5 and with an excess of reductant [3]. In the same rationale, Kaczerewska and co-authors suggested that it is possible to use novel gemini surfactants to synthetize greener silica mesoporous nanocapsules (SiNC) [2]. SiNC is a widely used nanomaterial that has been raising some environmental concerns due to the use of the cationic surfactant N-hexadecyl-N,N,N-trimethylammonium bromide (CTAB) that remains inside the nanostructure before being released over time when dispersed in seawater [5,9,10,13–16]. Thus, Kaczerewska et al. [2], used 1,4-bis-[N-(1-dodecyl)-N,N-dimethylammoniummethyl]benzene dibromide (QSB2-12) as a low toxicity template agent to replace CTAB [11]. Newly developed silica nanocapsules were quite similar to the conventional ones and exhibited significant reduction in the toxicity of such nanomaterials in marine microalgae and microcrustaceans [2].

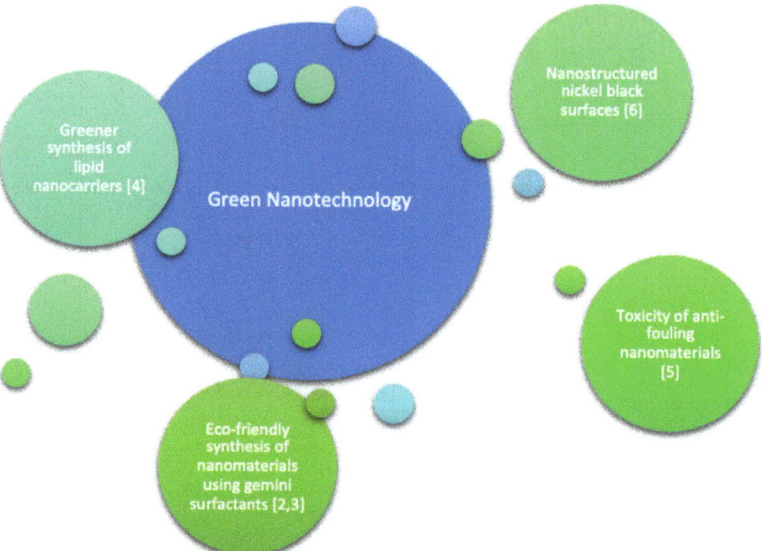

Figure 1. Main contributions of the present Special Issue to the field of green nanotechnology.

Another interesting topic is the development of safe and multi-purpose nanostructured lipid carriers, which have been widely proposed for pharmaceutical applications. The use of natural lipids is desirable for drug-delivery systems. Galvão et al. [4] assessed the influence of carvacrol in the crystallinity of solid natural lipids (stearic acid, beeswax and carnauba wax) to synthetize greener nanostructured lipid carriers. The authors showed that the higher the carvacrol content, the lower the crystallinity of the solid bulks of targeted lipids, demonstrating the promising properties of this monoterpenoid phenol towards the development of green drug delivery systems based on lipid nanoparticles [4].

Green nanotechnology has also been applied to the coatings industry through the replacement of toxic compounds [6], or their immobilization and controlled release over time [5,9,13]. Mennucci et al. [6] demonstrated that nanostructured nickel black surfaces can have good corrosion resistance and can be a great replacement for chromium finish, which is widely used in optical and solar applications, but also for decorative purposes. On the other hand, recent advances demonstrated that the nanoencapsulation of anti-fouling biocides (e.g., DCOIT, Zn and Cu pyrithiones within in SiNC or other engineered nanomaterials), widely used in maritime coatings, can significantly decrease their toxicity and hazard on marine species [9,10,13–16]. Santos et al. [5] demonstrated for the first time that the SiNC-DCOIT has high anti-fouling efficacy towards target early life stages of the tropical mussel *Perna perna*, while it is less toxic than free DCOIT during the larval development stage. This novel insight reinforces the benefits of the encapsulating toxic chemicals in nanocarriers.

3. Future Perspectives

Nowadays, science and technology are moving at a rapid pace and crossing scientific frontiers. Articles published in this Special Issue showed different directions for further progress in green nanotechnology. Future perspectives are dictated not only by new scientific ideas but largely by today's societal challenges, such as environmental regulations, and the need to increase innovation and sustainability in the industrial processes and decrease the loss of ecological biodiversity due to the combination of pollution and climate changes, among others, in the framework of the sustainable development goals (Agenda 2030) defined by the United Nations. As an example (Figure 2), and to avoid the repetition of past mistakes, the upcoming generation of nanomaterials must be truly environmentally friendly. For that purpose, synthesis should prioritize no/low toxic products, obtained from sustainable sources, and new nanomaterial must be carefully assessed in terms of environmental behavior, fate, effects, and hazard in the aquatic and terrestrial ecosystems, whenever possible. Efforts must be made to bridge the gap between industry and academia towards the development of green added-value and innovative nano-based solutions for real problems (e.g., corrosion, biofouling, water remediation, agrochemicals).

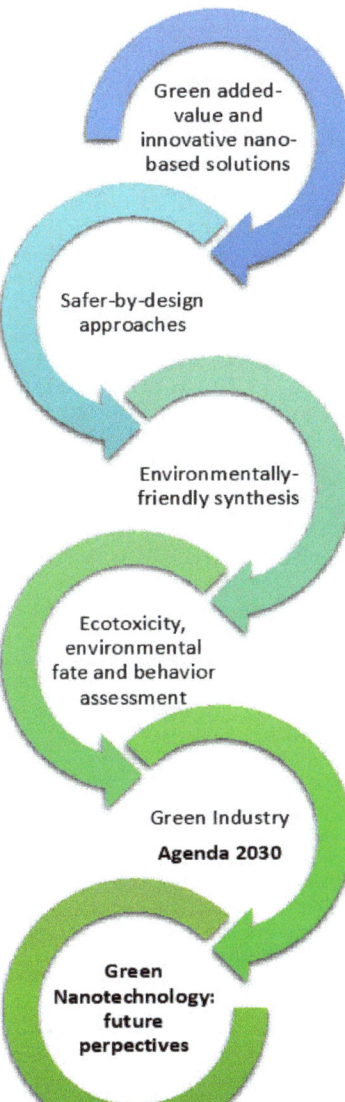

Figure 2. Future perspectives on the field of green nanotechnology.

Funding: R. Martins was hired under the Scientific Employment Stimulus—Individual Call (CEECIND/01329/2017), funded by national funds (OE), through FCT, in the scope of the framework contract foreseen in the numbers 4, 5 and 6 of the article 23, of the Decree-Law 57/2016, of 29 August, changed by Law 57/2017, of 19 July. O. Kaczerewska received funding from the European Union's Horizon 2020 research and innovation programme under the Marie Sklodowska-Curie grant agreement No 792945 (EcoGemCoat). This work was developed under the framework of the NANOGREEN project (CIRCNA/BRB/0291/2019), funded by the Portuguese Foundation for Science and Technology (FCT) through national funds (OE). This work was also carried out in the framework of SMARTAQUA project, which is funded by the Foundation for Science and Technology in Portugal (FCT), the Research Council of Norway (RCN-284002), Malta Council for Science and Technology (MCST), and co-funded by European Union's Horizon 2020 research and innovation program under the framework of ERA-NET Cofund MarTERA (Maritime and Marine Technologies for a new Era). Thanks are also due to FCT/MCTES for the financial support to CESAM (UIDP/50017/2020+UIDB/50017/2020) and CICECO-Aveiro Institute of Materials (UIDB/50011/2020; UIDP/50011/2020) through national funds.

Acknowledgments: We would like to thank all authors that contributed to the Special Issue which fostered the scientific excellence of it and the consequent book. A special thank is due to Wing Wang, SI Managing Editor, for all efforts, clarifications, and assistance during the invitation, preparation, and publication of the Special Issue.

Conflicts of Interest: The authors declare no conflict of interest.

References

1. Gottardo, S.; Mech, A.; Drbohlavová, J.; Małyska, A.; Bøwadt, S.; Riego Sintes, J.; Rauscher, H. Towards safe and sustainable innovation in nanotechnology: State-of-play for smart nanomaterials. *NanoImpact* **2021**, *21*, 100297. [CrossRef] [PubMed]
2. Kaczerewska, O.; Sousa, I.; Martins, R.; Figueiredo, J.; Loureiro, S.; Tedim, J. Gemini surfactant as a template agent for the synthesis of more eco-friendly silica nanocapsules. *Appl. Sci.* **2020**, *10*, 8085. [CrossRef]
3. Brycki, B.; Szulc, A.; Babkova, M. Synthesis of silver nanoparticles with gemini surfactants as efficient capping and stabilizing agents. *Appl. Sci.* **2021**, *11*, 154. [CrossRef]
4. Galvão, J.G.; Santos, R.L.; Lira, A.A.M.; Kaminski, R.; Sarmento, V.H.; Severino, P.; Dolabella, S.S.; Scher, R.; Souto, E.B.; Nunes, R.S. Stearic acid, beeswax and carnauba wax as green raw materials for the loading of carvacrol into nanostructured lipid carriers. *Appl. Sci.* **2020**, *10*, 6267. [CrossRef]
5. Santos, J.V.N.; Martins, R.; Fontes, M.K.; Galvao, B.; Silva, M.B.M.d.P.; Maia, F.; Abessa, D.M.d.S.; Perina, F.C. Can Encapsulation of the Biocide DCOIT Affect the Anti-Fouling Efficacy and Toxicity on Tropical Bivalves? *Appl. Sci.* **2020**, *10*, 8579. [CrossRef]
6. Mennucci, M.M.; Montes, R.; Bastos, A.C.; Monteiro, A.; Oliveira, P.; Tedim, J. Nanostructured nickel coating as replacement for black Cr (VI) finish. *Appl. Sci.* **2021**, *11*, 3924. [CrossRef]
7. Janczarek, M.; Endo-Kimura, M.; Wei, Z.; Bielan, Z.; Mogan, T.R.; Khedr, T.M.; Wang, K.; Markowska-Szczupak, A.; Kowalska, E. Novel Structures and Applications of Graphene-Based Semiconductor Photocatalysts: Faceted Particles, Photonic Crystals, Antimicrobial and Magnetic Properties. *Appl. Sci.* **2021**, *11*, 1982. [CrossRef]
8. Tiwari, E.; Singh, N.; Khandelwal, N.; Abdolahpur, F. Application of Zn/Al layered double hydroxides for the removal of nanoscale plastic debris from aqueous systems. *J. Hazard. Mater.* **2020**, *397*, 122769. [CrossRef] [PubMed]
9. Figueiredo, J.; Oliveira, T.; Ferreira, V.; Sushkova, A.; Silva, S.; Carneiro, D.; Cardoso, D.N.; Gonçalves, S.F.; Maia, F.; Rocha, C.; et al. Toxicity of innovative anti-fouling nano-based solutions to marine species. *Environ. Sci. Nano* **2019**, *6*. [CrossRef]
10. Figueiredo, J.; Loureiro, S.; Martins, R. Hazard of novel anti-fouling nanomaterials and biocides DCOIT and silver to marine organisms. *Environ. Sci. Nano* **2020**, *7*, 1670–1680. [CrossRef]
11. Kaczerewska, O.; Martins, R.; Figueiredo, J.; Loureiro, S.; Tedim, J. Environmental behaviour and ecotoxicity of cationic surfactants towards marine organisms. *J. Hazard. Mater.* **2020**, *392*, 122299. [CrossRef] [PubMed]
12. Gałuszka, A.; Migaszewski, Z.; Namieśnik, J. The 12 principles of green analytical chemistry and the SIGNIFICANCE mnemonic of green analytical practices. *TrAC Trends Anal. Chem.* **2013**, *50*, 78–84. [CrossRef]
13. Avelelas, F.; Martins, R.; Oliveira, T.; Maia, F.; Malheiro, E.; Soares, A.M.V.M.; Loureiro, S.; Tedim, J. Efficacy and ecotoxicity of novel anti-fouling nanomaterials in target and non-target marine species. *Mar. Biotechnol.* **2017**, *19*, 164–174. [CrossRef] [PubMed]
14. Gutner-Hoch, E.; Martins, R.; Oliveira, T.; Maia, F.; Soares, A.; Loureiro, S.; Piller, C.; Preiss, I.; Weis, M.; Larroze, S.; et al. Antimacrofouling efficacy of innovative inorganic nanomaterials loaded with booster biocides. *J. Mar. Sci. Eng.* **2018**, *6*, 6. [CrossRef]
15. Gutner-Hoch, E.; Martins, R.; Maia, F.; Oliveira, T.; Shpigel, M.; Weis, M.; Tedim, J.; Benayahu, Y. Toxicity of engineered micro- and nanomaterials with antifouling properties to the brine shrimp Artemia salina and embryonic stages of the sea urchin Paracentrotus lividus. *Environ. Pollut.* **2019**, *251*. [CrossRef] [PubMed]
16. De Jesus, É.P.S.; de Figueirêdo, L.P.; Maia, F.; Martins, R.; Nilin, J. Acute and chronic effects of innovative antifouling nanostructured biocides on a tropical marine microcrustacean. *Mar. Pollut. Bull.* **2021**, *164*. [CrossRef] [PubMed]

Article

Stearic Acid, Beeswax and Carnauba Wax as Green Raw Materials for the Loading of Carvacrol into Nanostructured Lipid Carriers

Juliana G. Galvão [1], Raquel L. Santos [1], Ana Amélia M. Lira [1], Renata Kaminski [2], Victor H. Sarmento [2], Patricia Severino [3,4,5], Silvio S. Dolabella [6], Ricardo Scher [6], Eliana B. Souto [7,8,*] and Rogéria S. Nunes [1,*]

[1] Department of Pharmacy, Federal University of Sergipe, São Cristóvão 49100-000, Brazil; julianaggalvao@gmail.com (J.G.G.); raquel.lines@hotmail.com (R.L.S.); ana_lira2@hotmail.com (A.A.M.L.)
[2] Department of Chemistry, Federal University of Sergipe, Itabaiana 49100-000, Brazil; re_kaminski@hotmail.com (R.K.); vhsarmento@gmail.com (V.H.S.)
[3] Laboratory of Nanotechnology and Nanomedicine (LNMED), Tiradentes University, Institute of Technology and Research (ITP), Aracaju 49010-390, Brazil; pattypharma@gmail.com
[4] Instituto de Tecnologia e Pesquisa (ITP), Av. Murilo Dantas, Aracaju 49032-490, Brazil
[5] Tiradentes Institute, Boston, MA 02125, USA
[6] Department of Morphology, Federal University of Sergipe, São Cristóvão 49100-000, Brazil; dolabellaufs@gmail.com (S.S.D.); rica.scher@gmail.com (R.S.)
[7] Department of Pharmaceutical Technology, Faculty of Pharmacy, University of Coimbra, 3000-548 Coimbra, Portugal
[8] CEB—Centre of Biological Engineering, University of Minho, Campus de Gualtar, 4710-057 Braga, Portugal
* Correspondence: ebsouto@ff.uc.pt (E.B.S.); rogeria.ufs@hotmail.com (R.S.N.)

Received: 13 August 2020; Accepted: 7 September 2020; Published: 9 September 2020

Abstract: The use of lipid nanoparticles as drug delivery systems has been growing over recent decades. Their biodegradable and biocompatible profile, capacity to prevent chemical degradation of loaded drugs/actives and controlled release for several administration routes are some of their advantages. Lipid nanoparticles are of particular interest for the loading of lipophilic compounds, as happens with essential oils. Several interesting properties, e.g., anti-microbial, antitumoral and antioxidant activities, are attributed to carvacrol, a monoterpenoid phenol present in the composition of essential oils of several species, including *Origanum vulgare*, *Thymus vulgaris*, *Nigella sativa* and *Origanum majorana*. As these essential oils have been proposed as the liquid lipid in the composition of nanostructured lipid carriers (NLCs), we aimed at evaluating the influence of carvacrol on the crystallinity profile of solid lipids commonly in use in the production of NLCs. Different ratios of solid lipid (stearic acid, beeswax or carnauba wax) and carvacrol were prepared, which were then subjected to thermal treatment to mimic the production of NLCs. The obtained binary mixtures were then characterized by thermogravimetry (TG), differential scanning calorimetry (DSC), small angle X-ray scattering (SAXS) and polarized light microscopy (PLM). The increased concentration of monoterpenoid in the mixtures resulted in an increase in the mass loss recorded by TG, together with a shift of the melting point recorded by DSC to lower temperatures, and the decrease in the enthalpy in comparison to the bulk solid lipids. The miscibility of carvacrol with the melted solid lipids was also confirmed by DSC in the tested concentration range. The increase in carvacrol content in the mixtures resulted in a decrease in the crystallinity of the solid bulks, as shown by SAXS and PLM. The decrease in the crystallinity of lipid matrices is postulated as an advantage to increase the loading capacity of these carriers. Carvacrol may thus be further exploited as liquid lipid in the composition of green NLCs for a range of pharmaceutical applications.

Keywords: carvacrol; stearic acid; beeswax; carnauba wax; nanostructured lipid carriers; crystallinity

1. Introduction

Carvacrol (or cymophenol) is chemically known as 5-isopropyl-2-methylphenol and is obtained from a range of aromatic plants such as oregano (*Origanum vulgare* L.), marjoram (*Origanum majorana* L.), black cumin (*Nigella sativa* L.) or thyme (*Thymus vulgaris* L.) [1,2]. Pharmacological properties, such as antioxidant [3,4], anti-inflammatory [5], analgesic [6], antitumor [7], antimicrobial [8] and antiprotozoal activities [9–11], have been attributed to this phenolic monoterpene. Its use in clinical settings is, however, compromised by its lipophilic character, resulting in low aqueous solubility, risk of oxidation and also high volatility [1]. The loading of carvacrol in lipid nanoparticles may be a promising strategy to reduce its volatility and improve its loading and bioavailability [12]. Only a limited number of studies have reported the loading of carvacrol into nanoparticles [13,14], while none has yet reported the use of nanostructured lipid carriers (NLCs) for this purpose.

It has already been demonstrated that lipid mixtures have a great impact on the chemical stability of volatile compounds [15–18]. Among lipid nanoparticles, NLCs are attractive colloidal carriers as they consist of nanosized particles based on a blend of solid and liquid lipids dispersed in an aqueous surfactant solution [19–22]. The combined ratio between solid and liquid lipid should ensure that the produced lipid matrix melts above 40 °C [20,23]. Besides being biodegradable, biocompatible and of low toxicity, the major advantages of NLCs include their green nature, capacity for protecting loaded drugs/actives from chemical degradation and the provision of a controlled release of the payload [19,24–29]. Due to their nanostructured matrix, obtained by disrupting the crystal packing of the solid lipid by mixing it with a liquid lipid, NLCs may also offer a triggered release [30]. The degrees of crystallization and polymorphism of the NLC matrices are strongly dependent on the ratio between the solid and liquid lipids [31]. A low degree of crystallinity of the matrix is usually associated with a higher payload. The use of lipid mixtures that crystallize in a less organized matrix is linked to higher loading capacity of the nanocarriers.

Several lipids excipients can be used for the production of NLCs, among which fatty acids (e.g., palmitic acid, stearic acid), fatty alcohols, mono-, di- and triglycerides, vegetable oils and waxes (e.g., carnauba wax, beeswax, cetyl palmitate) are the most frequently used [32–34]. In pharmaceutical products, the employment of natural lipids is desirable [35].

Stearic acid is a saturated 18-carbon chain fatty acid that melts around 69 °C and is obtained from both animal and vegetal sources. It shows higher in vivo tolerability and less toxicity than fats from synthetic origin [36,37]. Beeswax is a natural fatty material with a melting point ranging between 61 °C and 67 °C, obtained from the combs of bees (*Apis mellifera*) [38,39]. Carnauba wax is a plant exudate from the Brazilian "tree of life" (*Copernicia cerifera*), composed almost entirely of wax acid esters of C24 and C28 carboxylic acids and saturated long-chain mono-functional alcohols, that melts around 82 °C, thus showing high crystallinity [40]. Stearic acid, beeswax and carnauba wax have been selected as they are considered non-toxic and "generally recognized as safe" (GRAS) by the US Food and Drug Administration (FDA). The use of these three fats in the production of drug delivery systems has also been previously described [41,42].

Since the majority of methods used for the production of NLCs require heating, the volatility of carvacrol may compromise the loading capacity and encapsulation efficiency of the particles for the monoterpene. The liquid status of carvacrol may also affect the structure of the lipid matrix. The aim of this study was to evaluate the effect of carvacrol and its concentration on the crystallinity profile of lipid mixtures commonly used for the production of NLCs. Such lipid screening is commonly recommended prior to the development of NLCs, for the selection of the best lipid/lipid combination to achieve high payloads. Thermogravimetry (TG), differential scanning calorimetry (DSC), small-angle X-ray scattering (SAXS) and polarized light microscopy (PLM) were used for the physicochemical characterization of the lipid mixtures.

2. Material and Methods

2.1. Materials

Beeswax and carnauba wax were purchased from GM Ceras (São Paulo, Brazil). Stearic acid was obtained from Dinâmica® (Diadema, São Paulo, Brazil). Carvacrol (5-isopropyl-2-methylphenol, CAS Number 499-75-2) was purchased from Sigma-Aldrich® (St. Louis, MO, USA).

2.2. Preparation of the Binary Mixtures

Prior to the preparation of the binary mixture, the selected solid lipids, stearic acid (SA), beeswax (BW) and carnauba wax (CW), were first heated above their melting points, i.e., 58 °C, 63 °C and 82 °C, respectively, followed by cooling down to allow them to recrystallize [43]. To prepare the binary mixtures containing 10%, 25% and 50% of carvacrol, the solid lipids (SA, BW, CW) and carvacrol were melted at a temperature of 85 °C, were mechanically mixed for 5 minutes and then cooled down under continuous stirring until solidification. Composition of the binary mixtures is depicted in Table 1. The freshly prepared mixtures were then characterized.

Table 1. Composition of the binary mixtures (%, m/m).

Samples	Carvacrol (mg)	Stearic Acid (mg)	Beeswax (mg)	Carnauba Wax (mg)
SA 10%	0.10	0.90	—	—
SA 25%	0.25	0.75	—	—
SA 50%	0.50	0.50	—	—
BW 10%	0.10	—	0.90	—
BW 25%	0.25	—	0.75	—
BW 50%	0.50	—	0.50	—
CW 10%	0.10	—	—	0.90
CW 25%	0.25	—	—	0.75
CW 50%	0.50	—	—	0.50

2.3. Thermogravimetric (TG) Analysis

Thermogravimetric (TG) analysis was performed to evaluate the mass loss under heating, which was recorded in a Q50 TG (TA Instruments, New Castle, Delaware, USA). Samples of approximately 10 mg were heated up from 25 °C to 600 °C, applying a heating rate of 10 °C/min, under a dynamic argon atmosphere (50 mL/min) in platinum crucibles.

2.4. Differential Scanning Calorimetry (DSC) Analysis

Differential scanning calorimetry (DSC) analysis was performed to record the melting events and calculate the melting enthalpy using a DSC Q20 (TA Instruments, New Castle, Delaware, USA). Samples of approximately 3 mg were heated up from 25 °C to 100 °C, under a heating rate of 10 °C/min and dynamic argon atmosphere (50 mL/min) in aluminum crucibles.

2.5. Small-Angle X-ray Scattering (SAXS) Analysis

Small-angle X-ray scattering (SAXS) analysis was carried out to evaluate the polymorphism of the lipid mixtures. Diffractograms were obtained on the D1B-SAXS1 beamline at the Brazilian Synchrotron Light Laboratory (LNLS, Campinas, Brazil). Measurements were performed at room temperature, using a silicon-W/B4C toroidal multilayer mirror, collimated by a set of slits defining a pinhole geometry, at a wavelength of k = 1.499 Å and detected on a Pilatus 300 k detector (Dectris Ltd., Baden-Dättwil, Switzerland). The sample-to-detector distance was 814 mm, covering a scattering vector "q" ($q = (4\pi/\lambda)\sin\theta$) and ranging from 0.1 to 4.0 nm, where 2θ is the scattering angle. A standard silver behenate powder was measured to calibrate the sample-to-detector distance, detector tilt and direct beam position (at room temperature). From the total scattering intensity, the parasitic scattering produced was subtracted.

2.6. Polarized Light Microscopy (PLM) Analysis

The melted SA, BW, CW and their mixtures with carvacrol (Table 1) were examined under polarizing light with an Olympus model BX-51 microscope (Tokyo, Japan), coupled to a digital LC Color Evolution (PL-A662) camera. The software PixeLINK (Gloucester, Ontario, Canada) was used for recording the images. Briefly, the samples were heated above the melting point, placed between two glass plates and allowed to recrystallize by cooling down to room temperature. They were then analyzed under PLM at room temperature (25 °C). All samples were checked using 20× magnification.

3. Results and Discussion

For the development of stable NLC formulations, the evaluation of the degree of crystallinity and polymorphism of the lipid matrices is instrumental to ensure that the matrix remains in the solid state at room temperature. Besides, it is possible to predict polymorphic changes and the degree of miscibility between the solid lipid and the liquid lipid during recrystallization [35,44].

As for the NLC production, the methods usually require heating and the volatility of carvacrol compromises the success of its loading in the lipid matrices. TG analysis can be useful to determine the mass change after the tempering process. Figure 1 shows the mass change (%) of selected solid lipids (SA, BW and CW) and their respective binary mixtures containing 10, 25 and 50% of carvacrol, as a function of temperature.

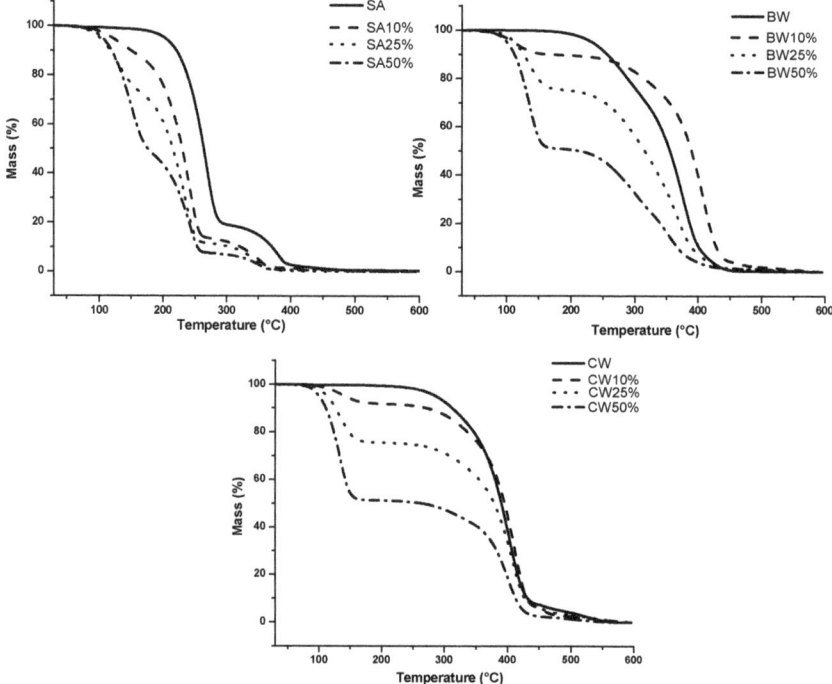

Figure 1. Thermogravimetric curves of the stearic acid (SA, upper left), beeswax (BW, upper right) and carnauba wax (CW, bottom) in comparison to their binary mixtures containing 10%, 25% and 50% of carvacrol (see Table 1 for composition).

Two mass loss events were identified in the differential thermogravimetric (DTG) curve of stearic acid (SA, Figure 1), the first being within the temperature range of 161 and 306 °C (Δm_1 = 79.63% T_{peak} DTG ~ 267 °C), which is typical of the SA decomposition, and the second within the range of 306 and 410 °C, attributed to the elimination of carbonaceous material [45,46]. The TG curves of beeswax (Figure 1, upper right) and carnauba wax (Figure 1, bottom) depict only one mass loss event within the temperature range of 180–480 °C (Δm_1 = 99.5% T_{peak} DTG ~ 396.7 °C) and 250–500 °C (Δm_1 = 98.5% T_{peak} DTG ~ 426.1 °C), respectively. The analysis of the binary mixtures shows the detection of carvacrol during the progressive mass loss with the increasing concentration of the oil and over the course of the experiment. It was interesting to see that the first mass loss event observed for all samples was within the range of the percentage of carvacrol (10%, 25% and 50%) in each of the binary mixtures (Table 2).

Table 2. Thermogravimetric data recorded for the first mass loss of the solid lipids (SA, BW and CW) containing 10%, 25% and 50% of carvacrol (see Table 1 for composition).

Samples	1st Loss in Mass Δm (%)
SA 10%	10.40%
SA 25%	24.44%
SA 50%	49.48%
BW 10%	9.66%
BW 25%	23.74%
BW 50%	47.84%
CW 10%	7.82%
CW 25%	24.20%
CW 50%	48.07%

Despite the volatility of the monoterpene [47], the results shown in Figure 1 demonstrate that when mixing carvacrol with the three solid lipids in the selected ratios, no mass change was recorded within the range of temperatures below 100 °C. This result assures the possibility of producing NLCs for the loading of carvacrol, since the production methods do not make use of temperatures above 100 °C. Carvacrol is an oil at room temperature (melting point of 1 °C, boiling point of 237.7 °C). No chemical degradation of carvacrol is thus anticipated from the thermal stress during the production of the melted mixtures.

DSC analysis provided information about the physical state of the melted lipids and their mixtures with carvacrol, degree of crystallinity, melting temperatures and respective enthalpies [48]. Figure 2 shows the DSC curves for the selected solid lipids (SA, BW and CW) and their binary mixtures containing 10%, 25% and 50% of carvacrol.

As seen in Figure 2 (left panel), SA showed an endothermic event between 40 °C and 65 °C (with the T_{peak} ~ 57 °C), which is related to the melting point of the solid lipid [45,46]. The melting point of beeswax was recorded at T_{peak} ~ 63 °C (middle panel) and that of carnauba wax was recorded at T_{peak} ~ 82 °C (right panel). The adding of monoterpene to the bulk lipids resulted in a slight shift down to lower temperatures, i.e., from 57 °C down to 56.34 °C, 54.9 °C and 50.2 °C, for stearic acid with 10%, 25% and 50% of carvacrol, respectively (Figure 2, left). Shifts in the melting points were also recorded for BW down to 61.4 °C, 58.5 °C and 50.9 °C, and for CW down to 81.36 °C, 77.3 °C and 72.9 °C, when added with 10%, 25% and 50% carvacrol, respectively.

The decrease in the enthalpy, which is the amount of heat involved in thermal events, was compared among the three solid lipids when increasing the concentration of carvacrol (Figure 3). When compared to pure solid lipids, the increase in the carvacrol content (10%, 25% and 50%) induced a decrease in the endothermic events related to melting and therefore the related enthalpy. Severino et al. studied the influence of the loading of capric/caprylic triglyceride mixtures (liquid lipid) in stearic acid matrices [37], demonstrating that the increase in melting enthalpy and crystallization was inversely proportional to the amount of oil in the formulation. A decrease in the crystallinity and melting enthalpy was observed, which translates a higher disorder of the lipid lattice [37]. Similar results were described when mixing theobroma oil and beeswax, contributing to the nanostructuring of the lipid matrix towards a greater disorder [35]. A less ordered matrix favors a higher loading capacity for active ingredients due to the increased number of voids and vacancies able to accommodate a higher number of molecules in the structure of the lattice. Due to the liquid character of carvacrol, its presence is likely to delay the complete crystalline rearrangement of the solid lipids.

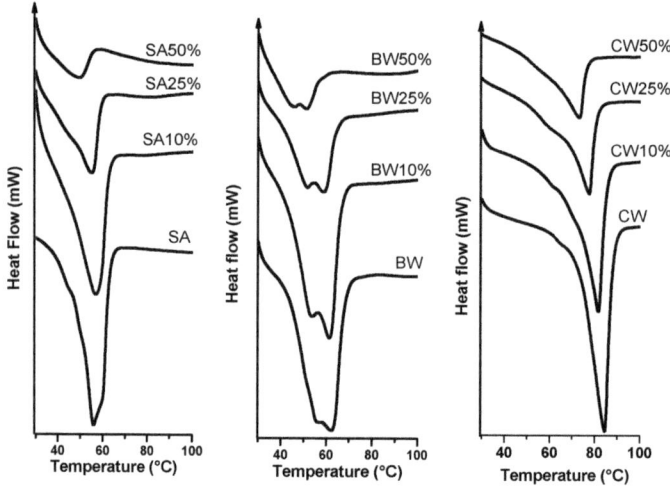

Figure 2. Differential scanning calorimetry curves of the stearic acid (left, SA), beeswax (middle, BW) and carnauba wax (right, CW), in comparison to their binary mixtures containing 10%, 25% and 50% of carvacrol (see Table 1 for composition).

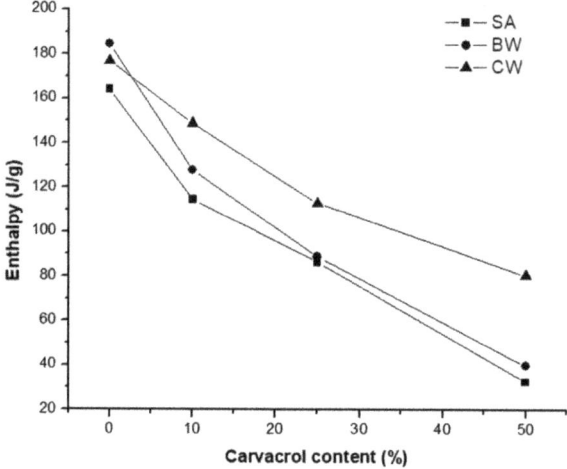

Figure 3. Variation of the enthalpy of stearic acid (SA), beeswax (BW) and carnauba wax (CW) with the increased concentration of carvacrol (%, w/w) in each binary mixture.

A depression in the melting point of the bulk solid lipids was detected upon the increased addition of carvacrol to all three solid lipids. The decrease in the melting peak (T_{peak}), onset temperature, enthalpy and the increase in the width of the melting event (WME) of SA, BW and CW when 10%, 25% and 50% of carvacrol was added to the solid lipids demonstrates that monoterpene is miscible in the tested concentration range with the three solid lipids (Table 3). Kasongo et al. [44] has reported similar results for the adding of Transcutol® HP up to 20% (w/w) to the solid lipid Precirol® ATO 5.

Table 3. Melting peak, onset temperature and width of the melting event of the bulk solid lipids and their binary mixtures with carvacrol obtained by differential scanning calorimetry (see Table 1 for composition).

Samples	Melting Peak (°C)	Onset (°C)	Width of Melting Event [1] (°C)
Bulk SA	57.0	37.0	28.0
SA 10%	56.5	36.0	29.0
SA 25%	54.9	34.3	30.0
SA 50%	50.2	34.0	30.0
Bulk BW	63.0	55.8	32.6
BW 10%	61.4	50.0	32.7
BW 25%	58.5	48.1	32.7
BW 50%	50.9	48.0	24.5
Bulk CW	82.0	39.5	35.5
CW 10%	81.4	38.6	36.5
CW 25%	77.3	38.2	37.1
CW 50%	72.9	36.2	39.2

[1] WME, width of the melting event, i.e., difference between endset and onset temperatures.

The same samples were also analyzed by SAXS, for the recording of the polymorphic changes of the bulk solid lipid with the addition of 10%, 25% and 50% of carvacrol (Figure 4).

Bulk SA showed five peaks at q (nm^{-1}) of 1.27, 1.53, 2.25, 3.06 and 3.84, which are typical of highly ordered materials (Figure 4, upper left). Both waxes (BW and CW) exhibited peaks with a periodicity, typical of lamellar structures (1:2:3:4 ...) [49]. With the increasing concentration of carvacrol, some peaks disappeared and/or had lower intensity of some of the solid lipid characteristic peaks, suggesting that monoterpene decreased the structure ordering of the bulks, confirming the DSC results [43]. Attama et al. also reported a lamellar crystal arrangement for beeswax [35].

Figure 5 shows the lattice spacing "d" as a function of carvacrol concentration [50]. Lattice spacing "d" was determined using the equation $d = 2\pi/q$. When increasing the concentration of carvacrol, the bulk SA did not show any changes in the lattice spacing (which remained d ~ 4 nm). For both waxes (BW and CW), the "d" value increased with the increase in monoterpene concentration. Bragg's equation ($2d \cdot \sin\theta = n\lambda$) was used to determine the interlayer distance (d) in the lipid lattice, where θ is the angle of diffraction, λ the wavelength and n the order of the crystalline plane [51]. When compared with the bulk lipid, the increase in the distance with the loading with a drug, anticipates the assumption that the drug molecules are within the lipid lattice. As shown in the patterns of Figure 5, the experimental lattice spacing of bulk stearic acid, beeswax and carnauba wax was, respectively, 4.15 nm, 7.25 nm and 8.50 nm. The mixing with increasing concentrations of carvacrol increased the interlayer spacing of beeswax and carnauba wax, but not of stearic acid. This result anticipates the assumption that both waxes would improve the loading capacity (LC) and encapsulation efficiency (EE) of carvacrol in NLCs composed of one of those lipids. The higher the concentration of carvacrol in the wax mixtures, the higher the lamellar spacing and thus the higher the LC and EE. According to Alexandridis et al., the amount of interface (and lamellae) decreases with the increase in the "d" value and, thus, increases the spacing between lamellae [50]. This result indicates that carvacrol is most likely to be placed in between the solid lipid lamellae of both waxes, increasing the "d" value, thus promoting a less ordered structure, which also confirms the DSC results. An amorphous polymorphic form (α-form) is associated with a less ordered structure of the lipid core, thus also improving both LC and EE [52].

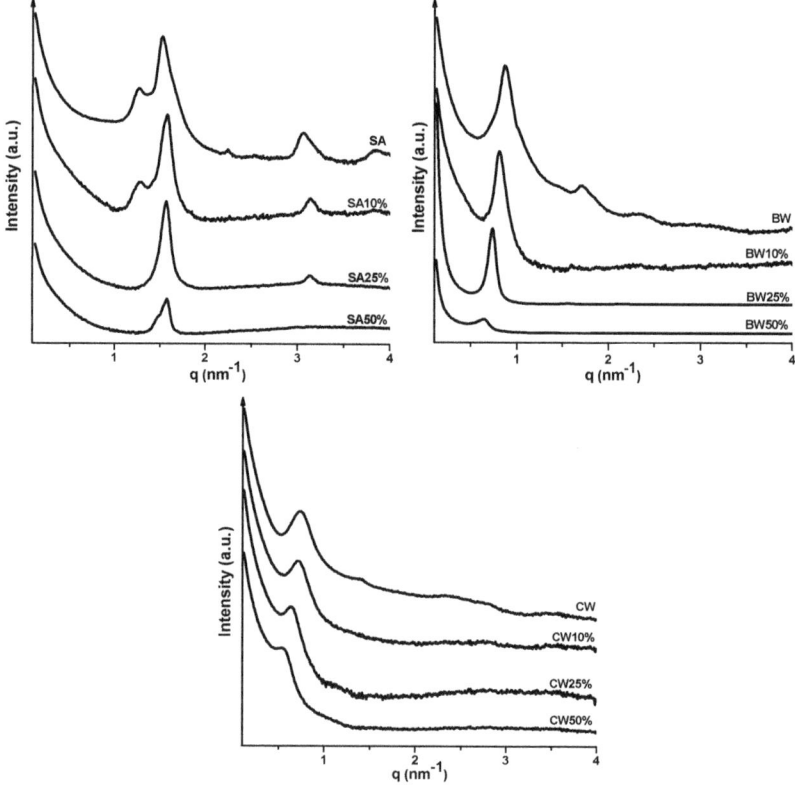

Figure 4. Scattering X-ray diffraction patterns $I(q)$ as a function of scattering vector (q) for stearic acid (SA, upper left), beeswax (BW, upper right), carnauba wax (CW, bottom) and their binary mixtures containing 10%, 25% and 50% of carvacrol (see Table 1 for composition).

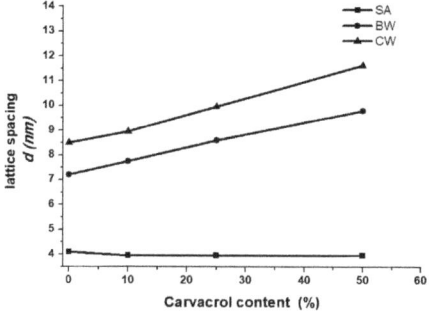

Figure 5. Value of lattice spacing "d" measured by small-angle X-ray scattering of bulk solid lipids (stearic acid, beeswax and carnauba wax) plotted as a function of carvacrol concentration (10%, 25% and 50%).

The optical behavior of solid lipids and their mixtures with carvacrol was checked by PLM. Materials can be classified either as anisotropic or as isotropic, depending on the effect that the material causes under polarized light. In the characterization of solid lipids, PLM can be used to observe microstructural changes [35].

Figure 6 shows the optical micrographs recorded at room temperature of bulk lipids (SA, BW and CW) and their binary mixtures with increased concentration of carvacrol (10%, 25% and 50%). The bulk solid lipids showed highly ordered crystalline microstructures, as previously demonstrated by DSC and SAXS analysis. SA exhibited a needle-shaped structure, while both waxes (BW and CW) clearly showed a maltese cross symbol of lamellar structure, confirming the results recorded with SAXS [43]. The addition of 10%, 25% and 50% of carvacrol resulted in a decrease in the size and thickness of these compared to the pure solid lipids (SA, BW and CW), suggesting a lower crystallinity. These results are in agreement with Gaillard et al., who showed that the crystallinity degree of beeswax decreased with an increase in the content of rosin [39].

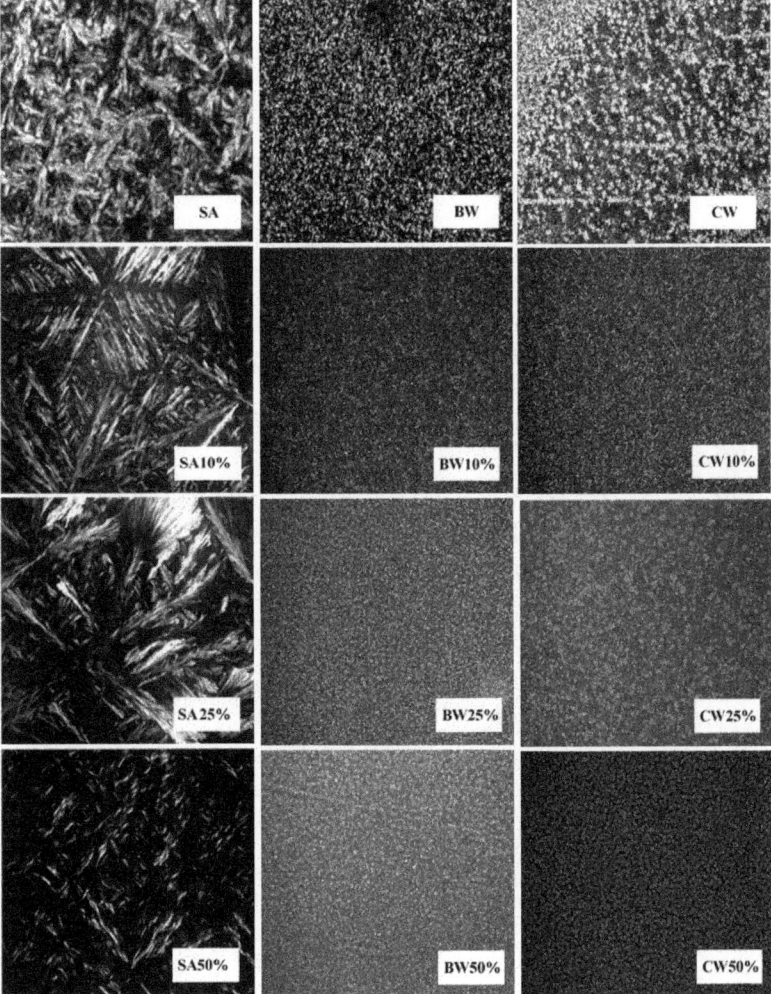

Figure 6. Optical micrographs produced with polarized light microscopy at room temperature corresponding to stearic acid, beeswax, carnauba wax and their binary mixtures containing 10%, 25% and 50% of carvacrol. Images obtained using 10 × magnification.

4. Conclusions

Due to the biological properties of carvacrol, this monoterpene has potential to be used as an active ingredient in pharmaceutical formulations; its oily liquid character makes this compound a suitable ingredient in formulating nanostructured lipid carriers (NLCs). Carvacrol was found to be well mixed with melted stearic acid, beeswax and carnauba wax, while the binary mixtures resulted in less ordered structures, which can be further exploited as drug delivery carriers. It is hypothesized that lipid mixtures containing 10%, 25% and 50% w/w of carvacrol in the solid lipids (SA, BW, and CW) can be used to obtain NLCs. Lower melting temperatures and enthalpy were recorded when adding the monoterpene to the three bulk lipids (confirmed by DSC). SAXS and PLM analyses demonstrated the presence of less ordered structures of the binary mixtures in comparison to the bulk counterparts. These binary mixtures can thus be explored in the production of NLCs for drug delivery.

Author Contributions: V.H.S., P.S., S.S.D., R.S. and R.S.N. contributed to the conceptualization, methodology, validation, formal analysis and investigation; J.G.G., R.L.S., A.A.M.L., R.K. and E.B.S. contributed to the writing—original draft preparation; E.B.S., P.S. and R.S.N. contributed to supervision, writing—review and editing, project administration, resources and funding acquisition. All authors have made a substantial contribution to the work. All authors have read and agreed to the published version of the manuscript.

Funding: This work was funded by the Conselho Nacional de Desenvolvimento Científico e Tecnológico (CNPq/Brazil), the Coordenação de Aperfeiçoamento de Pessoal de Nível Superior (CAPES/Brazil, Finance Code 001), by the Portuguese Science and Technology Foundation (FCT/MCT) and European Funds (PRODER/COMPETE) under the projects M-ERA-NET/0004/2015 and UIDB/04469/2020 (strategic fund), co-financed by FEDER, under the Partnership Agreement PT2020.

Acknowledgments: The authors wish to acknowledge the Brazilian Synchrotron Light Laboratory (LNLS) Campinas/SP for technical support in the SAXS measurements.

Conflicts of Interest: The authors declare no conflict of interest.

References

1. Santos, E.H.; Kamimura, J.A.; Hill, L.E.; Gomes, C.L. Characterization of carvacrol beta-cyclodextrin inclusion complexes as delivery systems for antibacterial and antioxidant applications. *LWT-Food Sci. Technol.* **2015**, *60*, 583–592. [CrossRef]
2. Silva, F.V.; Guimarães, A.G.; Silva, E.R.; Sousa-Neto, B.P.; Machado, F.D.; Quintans-Júnior, L.J.; Arcanjo, D.D.; Oliveira, F.A.; Oliveira, R.C. Anti-inflammatory and anti-ulcer activities of carvacrol, a monoterpene present in the essential oil of oregano. *J. Med. Food* **2012**, *15*, 984–991. [CrossRef] [PubMed]
3. Kumar, D.; Rawat, D.S. Synthesis and antioxidant activity of thymol and carvacrol based Schiff bases. *Bioorg. Med. Chem. Lett.* **2013**, *23*, 641–645.
4. Aeschbach, R.; Löliger, J.; Scott, B.; Murcia, A.; Butler, J.; Halliwell, B.; Aruoma, O. Antioxidant actions of thymol, carvacrol, 6-gingerol, zingerone and hydroxytyrosol. *Food Chem. Toxicol.* **1994**, *32*, 31–36. [CrossRef]
5. Guimarães, A.G.; Xavier, M.A.; de Santana, M.T.; Camargo, E.A.; Santos, C.A.; Brito, F.A.; Barreto, E.O.; Cavalcanti, S.C.; Antoniolli, Â.R.; Oliveira, R.C. Carvacrol attenuates mechanical hypernociception and inflammatory response. *Naunyn-Schmiedeberg's Arch. Pharmacol.* **2012**, *385*, 253–263.
6. De Sousa, D.P. Analgesic-like activity of essential oils constituents. *Molecules* **2011**, *16*, 2233–2252. [CrossRef]
7. Mueller, R.H.; Maeder, K.; Gohla, S. Solid lipid nanoparticles (SLN) for controlled drug delivery—A review of the state of the art. *Eur. J. Pharm. Biopharm.* **2000**, *50*, 161–177. [CrossRef]
8. Belda-Galbis, C.M.; Leufvén, A.; Martínez, A.; Rodrigo, D. Predictive microbiology quantification of the antimicrobial effect of carvacrol. *J. Food Eng.* **2014**, *141*, 37–43. [CrossRef]
9. De Amorim Santos, I.G.; Scher, R.; Rott, M.B.; Menezes, L.R.; Costa, E.V.; de Holanda Cavalcanti, S.C.; Blank, A.F.; dos Santos Aguiar, J.; da Silva, T.G.; Dolabella, S.S. Amebicidal activity of the essential oils of *Lippia* spp.(*Verbenaceae*) against *Acanthamoeba polyphaga* trophozoites. *Parasitol. Res.* **2016**, *115*, 535–540. [CrossRef]
10. De Melo, J.O.; Bitencourt, T.A.; Fachin, A.L.; Cruz, E.M.O.; de Jesus, H.C.R.; Alves, P.B.; de Fátima Arrigoni-Blank, M.; de Castro Franca, S.; Beleboni, R.O.; Fernandes, R.P.M. Antidermatophytic and antileishmanial activities of essential oils from *Lippia gracilis* Schauer genotypes. *Acta Trop.* **2013**, *128*, 110–115. [CrossRef]

11. Pastor, J.; García, M.; Steinbauer, S.; Setzer, W.N.; Scull, R.; Gille, L.; Monzote, L. Combinations of ascaridole, carvacrol, and caryophyllene oxide against Leishmania. *Acta Trop.* **2015**, *145*, 31–38. [CrossRef] [PubMed]
12. Bilia, A.R.; Piazzini, V.; Guccione, C.; Risaliti, L.; Asprea, M.; Capecchi, G.; Bergonzi, M.C. Improving on Nature: The Role of Nanomedicine in the Development of Clinical Natural Drugs. *Planta Med.* **2017**, *83*, 366–381. [CrossRef]
13. Amato, D.N.; Amato, D.V.; Mavrodi, O.V.; Braasch, D.A.; Walley, S.E.; Douglas, J.R.; Mavrodi, D.V.; Patton, D.L. Destruction of Opportunistic Pathogens via Polymer Nanoparticle-Mediated Release of Plant-Based Antimicrobial Payloads. *Adv. Healthc. Mater.* **2016**, *5*, 1094–1103. [CrossRef] [PubMed]
14. Keawchaoon, L.; Yoksan, R. Preparation, characterization and in vitro release study of carvacrol-loaded chitosan nanoparticles. *Colloids Surf. B Biointerfaces* **2011**, *84*, 163–171. [CrossRef]
15. Zielinska, A.; Martins-Gomes, C.; Ferreira, N.R.; Silva, A.M.; Nowak, I.; Souto, E.B. Anti-inflammatory and anti-cancer activity of citral: Optimization of citral-loaded solid lipid nanoparticles (SLN) using experimental factorial design and LUMiSizer(R). *Int. J. Pharm.* **2018**, *553*, 428–440. [CrossRef]
16. Carbone, C.; Martins-Gomes, C.; Caddeo, C.; Silva, A.M.; Musumeci, T.; Pignatello, R.; Puglisi, G.; Souto, E.B. Mediterranean essential oils as precious matrix components and active ingredients of lipid nanoparticles. *Int. J. Pharm.* **2018**, *548*, 217–226. [CrossRef] [PubMed]
17. Severino, P.; Andreani, T.; Chaud, M.V.; Benites, C.I.; Pinho, S.C.; Souto, E.B. Essential oils as active ingredients of lipid nanocarriers for chemotherapeutic use. *Curr. Pharm. Biotechnol.* **2015**, *16*, 365–370. [CrossRef]
18. Zielińska, A.; Ferreira, N.R.; Feliczak-Guzik, A.; Nowak, I.; Souto, E.B. Loading, release profile and accelerated stability assessment of monoterpenes-loaded solid lipid nanoparticles (SLN). *Pharm. Dev. Technol.* **2020**, *25*, 832–844. [CrossRef]
19. Vieira, R.; Severino, P.; Nalone, L.A.; Souto, S.B.; Silva, A.M.; Lucarini, M.; Durazzo, A.; Santini, A.; Souto, E.B. Sucupira Oil-Loaded Nanostructured Lipid Carriers (NLC): Lipid Screening, Factorial Design, Release Profile, and Cytotoxicity. *Molecules* **2020**, *25*, 685. [CrossRef]
20. Souto, E.B.; Baldim, I.; Oliveira, W.P.; Rao, R.; Yadav, N.; Gama, F.M.; Mahant, S. SLN and NLC for topical, dermal and transdermal drug delivery. *Expert Opin. Drug Deliv.* **2020**, *17*, 357–377. [CrossRef]
21. Sanchez-Lopez, E.; Espina, M.; Doktorovova, S.; Souto, E.B.; Garcia, M.L. Lipid nanoparticles (SLN, NLC): Overcoming the anatomical and physiological barriers of the eye-Part I-Barriers and determining factors in ocular delivery. *Eur. J. Pharm. Biopharm.* **2017**, *110*, 70–75. [CrossRef] [PubMed]
22. Sanchez-Lopez, E.; Espina, M.; Doktorovova, S.; Souto, E.B.; Garcia, M.L. Lipid nanoparticles (SLN, NLC): Overcoming the anatomical and physiological barriers of the eye-Part II-Ocular drug-loaded lipid nanoparticles. *Eur. J. Pharm. Biopharm.* **2017**, *110*, 58–69. [CrossRef]
23. Patidar, A.; Thakur, D.S.; Kumar, P.; Verma, J. A review on novel lipid based nanocarriers. *Int. J. Pharm. Pharm. Sci.* **2010**, *2*, 30–35.
24. Doktorovova, S.; Kovacevic, A.B.; Garcia, M.L.; Souto, E.B. Preclinical safety of solid lipid nanoparticles and nanostructured lipid carriers: Current evidence from in vitro and in vivo evaluation. *Eur. J. Pharm. Biopharm.* **2016**, *108*, 235–252. [CrossRef] [PubMed]
25. Doktorovova, S.; Souto, E.B.; Silva, A.M. Nanotoxicology applied to solid lipid nanoparticles and nanostructured lipid carriers—A systematic review of in vitro data. *Eur. J. Pharm. Biopharm.* **2014**, *87*, 1–18. [CrossRef]
26. Souto, E.B.; Zielinska, A.; Souto, S.B.; Durazzo, A.; Lucarini, M.; Santini, A.; Silva, A.M.; Atanasov, A.G.; Marques, C.; Andrade, L.N.; et al. (+)-Limonene 1,2-epoxide-loaded SLN: Evaluation of drug release, antioxidant activity and cytotoxicity in HaCaT cell line. *Int. J. Mol. Sci.* **2020**, *21*, 1449. [CrossRef]
27. Souto, E.B.; Souto, S.B.; Zielinska, A.; Durazzo, A.; Lucarini, M.; Santini, A.; Horbańczuk, O.K.; Atanasov, A.G.; Marques, C.; Andrade, L.N.; et al. Perillaldehyde 1,2-epoxide loaded SLN-tailored mAb: Production, physicochemical characterization and in vitro cytotoxicity profile in MCF-7 cell lines. *Pharmaceutics* **2020**, *12*, 161. [CrossRef]
28. Souto, E.B.; da Ana, R.; Souto, S.B.; Zielińska, A.; Marques, C.; Andrade, L.N.; Horbańczuk, O.K.; Atanasov, A.G.; Lucarini, M.; Durazzo, A.; et al. In Vitro Characterization, Modelling, and Antioxidant Properties of Polyphenon-60 from Green Tea in Eudragit S100-2 Chitosan Microspheres. *Nutrients* **2020**, *12*, 976. [CrossRef]
29. Silva, A.M.; Martins-Gomes, C.; Fangueiro, J.F.; Andreani, T.; Souto, E.B. Comparison of antiproliferative effect of epigallocatechin gallate when loaded into cationic solid lipid nanoparticles against different cell lines. *Pharm. Dev. Technol.* **2019**, *24*, 1243–1249. [CrossRef]

30. Zheng, M.; Falkeborg, M.; Zheng, Y.; Yang, T.; Xu, X. Formulation and characterization of nanostructured lipid carriers containing a mixed lipids core. *Colloids Surf. A Physicochem. Eng. Asp.* **2013**, *430*, 76–84. [CrossRef]
31. Yang, Y.; Corona, A.; Schubert, B.; Reeder, R.; Henson, M.A. The effect of oil type on the aggregation stability of nanostructured lipid carriers. *J. Colloid Interface Sci.* **2014**, *418*, 261–272. [CrossRef] [PubMed]
32. Rosiaux, Y.; Jannin, V.; Hughes, S.; Marchaud, D. Solid lipid excipients—Matrix agents for sustained drug delivery. *J. Control. Release* **2014**, *188*, 18–30. [CrossRef]
33. Souto, E.B.; Almeida, A.J.; Müller, R.H. Lipid Nanoparticles (SLN®, NLC®) for Cutaneous Drug Delivery: Structure, Protection and Skin Effects. *J. Biomed. Nanotechnol.* **2007**, *3*, 317–331. [CrossRef]
34. Souto, E.B.; Doktorovova, S. Chapter 6—Solid lipid nanoparticle formulations pharmacokinetic and biopharmaceutical aspects in drug delivery. *Methods Enzym.* **2009**, *464*, 105–129. [CrossRef]
35. Attama, A.; Schicke, B.; Müller-Goymann, C. Further characterization of theobroma oil–beeswax admixtures as lipid matrices for improved drug delivery systems. *Eur. J. Pharm. Biopharm.* **2006**, *64*, 294–306. [CrossRef]
36. Fundarò, A.; Cavalli, R.; Bargoni, A.; Vighetto, D.; Zara, G.P.; Gasco, M.R. Non-stealth and stealth solid lipid nanoparticles (SLN) carrying doxorubicin: Pharmacokinetics and tissue distribution after iv administration to rats. *Pharmacol. Res.* **2000**, *42*, 337–343. [CrossRef] [PubMed]
37. Severino, P.; Pinho, S.C.; Souto, E.B.; Santana, M.H. Polymorphism, crystallinity and hydrophilic–lipophilic balance of stearic acid and stearic acid–capric/caprylic triglyceride matrices for production of stable nanoparticles. *Colloids Surf. B Biointerfaces* **2011**, *86*, 125–130. [CrossRef]
38. Attama, A.A.; Müller-Goymann, C.C. Effect of beeswax modification on the lipid matrix and solid lipid nanoparticle crystallinity. *Colloids Surf. A Physicochem. Eng. Asp.* **2008**, *315*, 189–195. [CrossRef]
39. Gaillard, Y.; Mija, A.; Burr, A.; Darque-Ceretti, E.; Felder, E.; Sbirrazzuoli, N. Green material composites from renewable resources: Polymorphic transitions and phase diagram of beeswax/rosin resin. *Thermochim. Acta* **2011**, *521*, 90–97. [CrossRef]
40. Talens, P.; Krochta, J.M. Plasticizing effects of beeswax and carnauba wax on tensile and water vapor permeability properties of whey protein films. *J. Food Sci.* **2005**, *70*, E239–E243. [CrossRef]
41. Baek, J.-S.; So, J.-W.; Shin, S.-C.; Cho, C.-W. Solid lipid nanoparticles of paclitaxel strengthened by hydroxypropyl-β-cyclodextrin as an oral delivery system. *Int. J. Mol. Med.* **2012**, *30*, 953–959. [CrossRef] [PubMed]
42. Kheradmandnia, S.; Vasheghani-Farahani, E.; Nosrati, M.; Atyabi, F. Preparation and characterization of ketoprofen-loaded solid lipid nanoparticles made from beeswax and carnauba wax. *Nanomed. Nanotechnol. Biol. Med.* **2010**, *6*, 753–759. [CrossRef] [PubMed]
43. Galvao, J.G.; Trindade, G.G.; Santos, A.J.; Santos, R.L.; Chaves Filho, A.B.; Lira, A.A.M.; Miyamoto, S.; Nunes, R.S. Effect of Ouratea sp. butter in the crystallinity of solid lipids used in nanostructured lipid carriers (NLCs). *J. Therm. Anal. Calorim.* **2016**, *123*, 941–948. [CrossRef]
44. Kasongo, K.W.; Pardeike, J.; Müller, R.H.; Walker, R.B. Selection and characterization of suitable lipid excipients for use in the manufacture of didanosine-loaded solid lipid nanoparticles and nanostructured lipid carriers. *J. Pharm. Sci.* **2011**, *100*, 5185–5196. [CrossRef] [PubMed]
45. Chen, Z.; Cao, L.; Shan, F.; Fang, G. Preparation and characteristics of microencapsulated stearic acid as composite thermal energy storage material in buildings. *Energy Build.* **2013**, *62*, 469–474. [CrossRef]
46. Haywood, A.; Glass, B.D. Pharmaceutical excipients—Where do we begin. *Aust. Prescr.* **2011**, *34*, 112–114. [CrossRef]
47. Higueras, L.; López-Carballo, G.; Gavara, R.; Hernández-Muñoz, P. Incorporation of hydroxypropyl-β-cyclodextrins into chitosan films to tailor loading capacity for active aroma compound carvacrol. *Food Hydrocoll.* **2015**, *43*, 603–611. [CrossRef]
48. Chorilli, M.; Campos, G.R.; Bolfarini, P.M. Desenvolvimento e estudo da estabilidade físico-química de emulsões múltiplas A/O/AEO/A/O acrescidas de filtros químicos e manteiga de karité. *Lat. Am. J. Pharm.* **2009**, *28*, 936–940.
49. Martins, A.J.; Cerqueira, M.A.; Fasolin, L.H.; Cunha, R.L.; Vicente, A.A. Beeswax organogels: Influence of gelator concentration and oil type in the gelation process. *Food Res. Int.* **2016**, *84*, 170–179. [CrossRef]
50. Alexandridis, P.; Olsson, U.; Lindman, B. A record nine different phases (four cubic, two hexagonal, and one lamellar lyotropic liquid crystalline and two micellar solutions) in a ternary isothermal system of an amphiphilic block copolymer and selective solvents (water and oil). *Langmuir* **1998**, *14*, 2627–2638. [CrossRef]

51. Shah, R.M.; Bryant, G.; Taylor, M.; Eldridge, D.S.; Palombo, E.A.; Harding, I.H. Structure of solid lipid nanoparticles produced by a microwave-assisted microemulsion technique. *RSC Adv.* **2016**, *6*, 36803–36810. [CrossRef]
52. De Souza, A.L.R.; Andreani, T.; Nunes, F.M.; Cassimiro, D.L.; de Almeida, A.E.; Ribeiro, C.A.; Sarmento, V.H.V.; Gremião, M.P.D.; Silva, A.M.; Souto, E.B. Loading of praziquantel in the crystal lattice of solid lipid nanoparticles. *J. Therm. Anal. Calorim.* **2012**, *108*, 353–360. [CrossRef]

© 2020 by the authors. Licensee MDPI, Basel, Switzerland. This article is an open access article distributed under the terms and conditions of the Creative Commons Attribution (CC BY) license (http://creativecommons.org/licenses/by/4.0/).

Article

Nanostructured Black Nickel Coating as Replacement for Black Cr(VI) Finish

Marina M. Mennucci [1,*], Rodrigo Montes [1], Alexandre C. Bastos [1], Alcino Monteiro [2], Pedro Oliveira [2], João Tedim [1] and Mário G. S. Ferreira [1]

[1] DEMaC—Department of Materials and Ceramic Engineering, CICECO—Aveiro Institute of Materials, University of Aveiro, 3810-193 Aveiro, Portugal; rodrigomontes@ua.pt (R.M.); acbastos@ua.pt (A.C.B.); joao.tedim@ua.pt (J.T.); mgferreira@ua.pt (M.G.S.F.)

[2] Leica—Aparelhos Ópticos de Precisão, S.A., Rua da Leica 55, 4760-810 Lousado VNF, Portugal; amonteiro@leica.pt (A.M.); poliveira@leica.pt (P.O.)

* Correspondence: marinamennucci@ua.pt

Featured Application: The black coatings described in this work find application in decorative finishes, optical instruments, and photovoltaic modules.

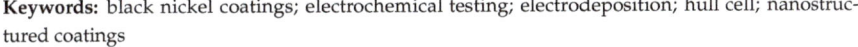

Abstract: This work compares different electrodeposition procedures to produce nickel black coatings as greener and less toxic alternatives to Cr(VI)-based coatings used in different applications. Nickel and nickel-plated brass served as substrates in studies with a Hull cell and polarization curves. After a set of comparative experiments, the best electrodeposition procedure was further studied and optimized. Optimal conditions were found with a bath consisting of 75 g/L $NiCl_2 \cdot 6H_2O$ + 30 g/L NaCl and a current density of 0.143 A dm^{-2} applied for 5 min at room temperature. Furthermore, a pre-treatment with 18.5 vol.% of hydrochloric acid in water was found to be necessary to warrant good coating adhesion to the substrate. The black color is attributed to the development of a nanostructured surface that absorbs the incident light. Corrosion testing was performed in 0.5 M NaCl aqueous solution using electrochemical impedance spectroscopy (EIS) and polarization tests.

Keywords: black nickel coatings; electrochemical testing; electrodeposition; hull cell; nanostructured coatings

Citation: Mennucci, M.M.; Montes, R.; Bastos, A.C.; Monteiro, A.; Oliveira, P.; Tedim, J.; Ferreira, M.G.S. Nanostructured Black Nickel Coating as Replacement for Black Cr(VI) Finish. *Appl. Sci.* **2021**, *11*, 3924. https://doi.org/10.3390/app11093924

Academic Editors: Alberto Milani, Roberto Martins and Olga Barbara Kaczerewska

Received: 30 March 2021
Accepted: 22 April 2021
Published: 26 April 2021

Publisher's Note: MDPI stays neutral with regard to jurisdictional claims in published maps and institutional affiliations.

Copyright: © 2021 by the authors. Licensee MDPI, Basel, Switzerland. This article is an open access article distributed under the terms and conditions of the Creative Commons Attribution (CC BY) license (https://creativecommons.org/licenses/by/4.0/).

1. Introduction

Black coatings are surface finishes with functional [1–10] or decorative [4,7–9] purposes. The films must satisfy characteristics such as adhesion, mechanical, and corrosion resistance. High absorbance capacity is required for solar [9] and optical applications [11,12], while decorative finishes must be resistant to wear and preserve a good and uniform appearance throughout the service life. Often these coatings are obtained by electrodeposition from hexavalent chromium baths [4,9,13–15]. Despite the good quality of chromium coatings, both health and environmental concerns associated with chromium (VI)-derived species [16–20] led to the prohibition of hexavalent chromium products [19,21] thereby forcing the industry to search for greener and safer alternatives [2,5–11,20]. The risks associated with chromium (VI) compounds [21–25] include allergic skin reactions that appear immediately after contact, regardless of the dose. Brief exposure to high concentrations can result in ulceration of the exposed skin, perforation of the respiratory tract, and irritation of the intestinal tract [16–18,21,25]. Kidney and liver damage have also been reported [16–18]. In addition, chromium (VI) compounds are classified as carcinogenic [16–18,20,21]. Prolonged occupational exposure to air with chromium levels higher than those present in the natural environment has been linked to lung cancer [16–18,21]. Workers in the chromium compounds manufacturing industry are those most at risk, but problems appear also in workers from the electroplating, metal artifacts handling, chrome alloys and stainless-steel welding, and chromium pigments industries.

Greener and less toxic baths for electroplating black coatings based on nickel [1–10], trivalent chromium [9,26], zinc [9,27], phosphate [9,28] and molybdenum [9,29] have been proposed. Some of these coatings are nanostructured [1,30,31]. Other procedures to deposit black coatings include electroless deposition [9,32] and vapor phase deposition [9,33,34].

Lira-Cantú et al. [1] studied the electrochemical deposition of black nickel solar absorber coatings on stainless steel AISI316L for thermal solar cells. They electrodeposited a bright nickel coating followed by a black nickel solar absorber coating and a top antireflection coating based on TEOS (tetraethyl orthosilicate). The bath solution was nickel chloride 75 g/L and sodium chloride 30 g/L, at room temperature (25–27 °C). To obtain the solar absorber coating, it was necessary to apply two different current densities (1.4 mA/cm^2 and 2.6 mA/cm^2) with an intermediate drying step with nitrogen. A small modification in current density did not affect solar absorptance, but deposition time did. Times greater than 60 s and 90 s for the first and second depositions, respectively, led to high solar absorptance but poor thermal emittance. The thickness of the black nickel coating varied between 1.3 and 1.7 µm, depending on the deposition time. The TEOS coating improved the absorptivity value when dried at 200 °C, and decreased it when dried at 300 °C.

With the same bath composition, Wäckelgård [2] varied the deposition parameters to obtain maximum solar absorptance and minimum thermal emittance from coatings formed on aluminum or copper substrates. As aluminum is difficult to plate directly, two steps were carried out prior to the black coating deposition. First, a pretreatment with double-zincate immersion technique was applied [3], followed by a bright nickel film deposition. Two temperatures were tested, 19 °C and 25 °C, under two current regimes: 70 mA/dm^2 for 2–6 min + 130 mA/dm^2 for 1–2 min in one case, and in another 110 mA/dm^2 for 3 min + 220 mA/dm^2 for 1 min. According to Wäckelgård [2], to achieve high solar selectivity two distinct sublayers must be present and for that, the process must be interrupted, with the sample removed from the solution and the surface dried before the second plating stage. If the sample remains in the bath, no distinct separation between the two layers is produced by the different current regimes. The top coating consisted of flakes about 200 nm wide and 10 nm thick.

Dennis and Such [4] explained that black nickel coatings have little abrasion and corrosion resistance and to improve them the deposition of the black coating must be made over an undercoat of dull or bright nickel. They also proposed two baths for producing black nickel coatings, one based on nickel sulfate and the other on nickel chloride. Both had a high amount of zinc and thiocyanate ions, which were believed to be responsible for the black color.

Ibrahim [5] modified the Watts bath by adding KNO_3 and varying its concentration. The best procedure for making a highly adherent black nickel coating on steel was achieved using a solution of 0.63 M $NiSO_4 \cdot 6H_2O$, 0.09 M $NiCl_2 \cdot 6H_2O$, 0.3 M H_3BO_3, and 0.2 M KNO_3 at pH = 4.6, i = 0.5A/dm^2, T = 25 °C, and t = 10min. Instantaneous nucleation was indicated by potentiostatic current-time transients. The black nickel coating consisted of metallic nickel with Ni(111) preferred orientation, as indicated by XRD studies.

Karuppiah et al. [6] produced a nickel-cobalt black coating on a copper panel precoated with a 10 µm thick nickel layer applied using the Watts bath. The plating solution contained 10 g/L of nickel sulfate, 10 g/L of cobalt sulfate, and 10 g/L of ammonium acetate. The solution pH was 6.2 at 308 K, and the current density varied between 3 to 10 A/dm^2, a range obtained from Hull cell studies. Scanning electron microscopy images revealed that the deposit was constituted by particles of different sizes and shapes. Due to optical interference and surface roughness, the deposit had a high degree of solar absorption. The values of solar absorptance and thermal emittance of the coating were influenced by the deposition time and the current density.

Cu-Ni black coatings were deposited by Aravinda et al. [7,8] on molybdenum, either from an ethylenediaminetetraacetic (EDTA) complex bath solution [7] or from a triethanolamine (TEA) complex bath solution containing ammonium persulfate (AP) [8].

The deposits made from the EDTA solution or from the TEA solution were black, uniform, pore-free, and demonstrated good solar selectivity.

Li et al. [31] studied the electrodeposition of nanostructured black nickel thin films on brass. The plating solution was 100 g/L $NiSO_4·6H_2O$ + 40 g/L $NiCl_2·6H_2O$ + 30 g/L H_3BO_3. Several parameters were analyzed in this work: current density (1–5 mA/cm^2), temperature (20–80 °C), pH (2–6), stirring speed (200–1200 rpm), and electrodeposition time (10–60 min). All parameters influenced the color of the coating, which varied between white, gray, and black, depending on the deposition conditions [31]. Varying the pH and keeping fixed all the other parameters (60 °C, 3.0 mA/cm^2, t = 30 min, stirring speed = 900 rpm), the black color appeared with pH = 2–3, gray with pH = 4, and white with pH = 5–6. When only the current density was varied (the fixed parameters were 60 °C, pH = 3, t = 30 min, stirring speed = 900 rpm), the black deposit appeared with a current density of 1–3 mA/cm^2, gray with 4 mA/cm^2, and white between 5 and 6 mA/cm^2. When the temperature was varied, black deposits appeared only in the 60–80 °C range. The range of stirring speed to obtain a black deposit was between 900 and 1000 rpm [31]. It was also verified that the black nickel film possessed a nanostructure with an average grain diameter of about 50 nm. More black coatings have been reviewed by Takadoum [9].

The present work compares electrodeposition procedures described in the literature for the production of black coatings using baths with nickel as the main metallic constituent. The objective is to select the best deposition procedure to produce black coatings on nickel substrates for decorative applications.

2. Materials and Methods

2.1. Materials, Reagents, and Experimental Parameters

A pure nickel metal sheet (99.5% purity, 2.0 mm thick, Alfa Aesar/Germany) and nickel-plated brass samples (Leica/Portugal) were used as metal substrates. The nickel-based electroplating baths studied in this work are presented in Table 1 with the respective references and were prepared with pro-analysis grade reagents and distilled water.

Table 1. Deposition baths and experimental conditions for producing black nickel coatings (adapted from [9]).

Procedure	1	2	3	4	5	6
Reagents and Parameters	Quantity (g/L)	Quantity (g/L)	Quantity (M)	Quantity (g/L)	Quantity (g/L)	Quantity (g/L)
$NiSO_4·6H_2O$			0.63	10		20
$NiCl_2·6H_2O$	75	75	0.09			
H_3BO_3			0.3		10	10
KNO_3			0.2			
NaCl	30					
NH_4Cl		30				
$ZnCl_2$		30				
NaSCN		15				
$CoSO_4$				10		
CH_3COONH_4				10		
EDTA					2.5	
$(NH_4)_2Ni(SO_4)_2·6H_2O$					20	
$Cu(NO_3)_2$					2.5	
$CuSO_4·5H_2O$						20
$(NH_4)_2S_2O_8$						10
$(HOCH_2CH_2)_3N$						20 mL/L
pH		3.5–5.5	4.6	6.2	5.2	5
Temperature (°C)	25–27	Room	25	35	27	27
Current density (A/dm^2)	0.14 (90 s) + 0.26 (60 s)	0.15	0.5	3–10	0.5	0.5
Duration		30 min	10 min	5–50 s	30 s	30 s
Reference	[1]	[4]	[5]	[6]	[7]	[8]

2.2. Electrodeposition

The black coating deposition took place in 10.5 × 15.5 × 10 cm^3 polyethylene containers with nickel-plated brass samples as cathodes (circular shape, 3 cm diameter, 1 mm thick sheet) and pure nickel anodes (anode/cathode area ratio of 1/1.5), using a Keysight N 5751A (USA) power supply and the experimental conditions presented in Table 1. Tests were also carried out in a Hull cell (Figure 1) [35,36] with a pure Ni anode and a pure nickel plate (0.2 × 10 × 7 cm^3) as cathode, following the DIN 50,957 standard [35]. The coatings adhesion to the substrate was evaluated with the cross-cut test ISO 2409:2007-08 [37]. The coating surface morphology was characterized by scanning electron microscopy using a Hitachi SU70 (Japan) microscope with an electron beam energy of 15 or 25 keV.

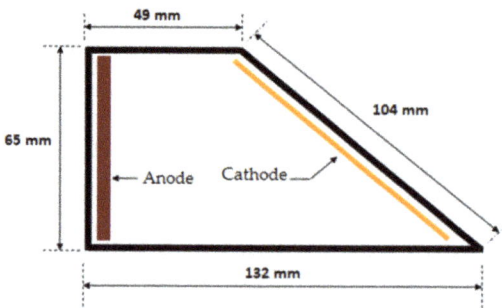

Figure 1. Top view of the 250 mL Hull cell used in this study.

The electrodeposition process was further investigated by polarization curves measured with an Autolab PGSTAT204 potentiostat (The Netherlands) controlled by the Nova 2.1.4 software, in an electrochemical cell with a three-electrode arrangement, comprising 1 cm^2 pure Ni samples as working electrodes, a RedRod reference electrode (Radiometer Analytical REF201/Villeurbanne Cedex—France), 199 mV vs. SHE at 25 °C) and a platinum wire as the auxiliary electrode. The polarization curves were obtained by sweeping the potential from +0.1 V vs. OCP (open circuit potential) down to −1.3 V vs. Red Rod at a scan rate of 0.5 mV/s. The polarization tests were carried out in triplicate to evaluate the reproducibility of the tests.

2.3. Corrosion Testing

The corrosion resistance was studied by electrochemical techniques. The electrochemical cell was made by gluing polymethylmethacrylate tubes to the sample surface with epoxy adhesive. This delimited a working electrode area of 2 cm^2 at the bottom of the tube and the reference and auxiliary electrodes were placed inside the tube. The testing solution was 0.5 M NaCl. Electrochemical impedance spectroscopy (EIS) diagrams were obtained with a Gamry Reference 600 (USA) in potentiostatic mode at OCP, applying an ac perturbation amplitude of 10 mV (rms) in the frequency range 100 kHz to 1 mHz, with 10 points per decade. The reference electrode was the RedRod from Radiometer (Villeurbanne Cedex, France) and the auxiliary electrode was a coiled platinum wire. Polarization curves were measured with the same electrochemical cell arrangement but using an Autolab PGSTAT204 (Utrecht, The Netherlands) potentiostat and a saturated calomel electrode (SCE—Radiometer Analytical/ Villeurbanne Cedex—France) as reference. The potential sweeps were performed in the anodic direction, from −0.1 V vs. OCP to +1.0 V$_{SCE}$, at a scan rate of 1 mV/s.

3. Results and Discussion

3.1. Coatings Produced According to the Experimental Conditions Described in the Literature

As stated above, the objective of this work was to select a procedure to produce a nickel black coating for decorative applications that could replace the Cr(VI) traditional one. The first step was to compare the coatings produced by directly following the procedures reported in a selection of references from the literature [1–9], as depicted in Table 1. Figure 2 shows optical images of the samples after electrodeposition. The sample presented in Figure 2a shows the substrate before deposition. Figure 2b corresponds to the sample treated according to procedure 1, in which a black matte coating was deposited. Samples presented in Figure 2c, Figure 2d were obtained following procedures 2 and 3 respectively, but in these cases, the deposition attempts were not successful. On the other hand, procedure 4 produced pulverulent black deposits with current densities of 5 A/dm^2 (Figure 2e) and 3.57 A/dm^2 (Figure 2f). No deposition was observed with procedure 5 (Figure 2g) while procedure 6 produced a non-uniform grayish to brownish deposit (Figure 2h). The discrepant results could be a consequence of using a substrate and/or cell geometry different from those established in the reference works. A direct application of the original experimental parameters is unlikely to be successful. It is evident that the experimental parameters must be modified to produce a black coating and then optimized for obtaining the desired finish quality.

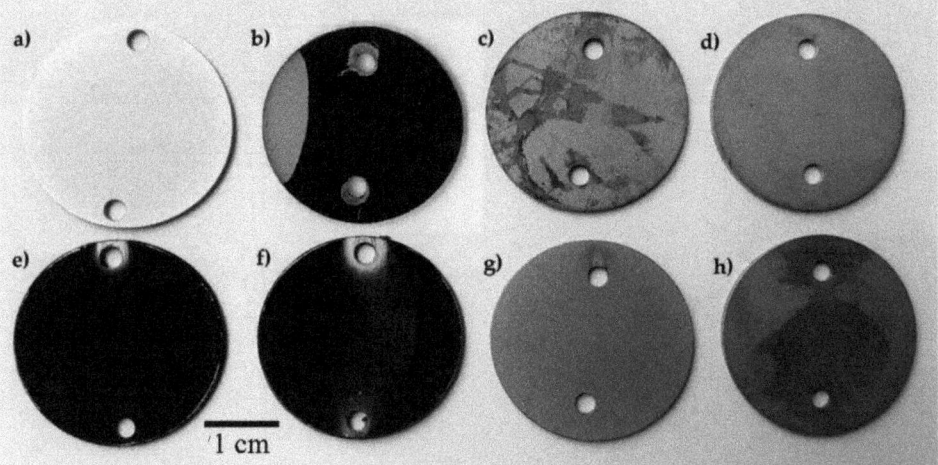

Figure 2. Nickel-plated brass samples after black coating deposition following the procedures in Table 1. (**a**) sample before deposition (substrate), (**b**) procedure 1, (**c**) procedure 2, (**d**) procedure 3, (**e**,**f**) procedure 4, (**g**) procedure 5, (**h**) procedure 6.

3.2. Tests Using the Hull Cell

Among the several parameters that influence coating deposition, the current density is probably the most determinant. In this work, a Hull cell was used to find the optimal range of current density for the formation of a black deposit. This small-scale electrodeposition tank has a trapezoidal shape to create a varying distance between the anode and cathode—Figure 1—thus making it possible to analyze a wide range of current densities in a single experiment. The Hull cell allows studying the relationship between current density and the quality of the deposited metallic layer varying the plating conditions, as for example bath constituents, temperature, agitation, among others [14,15]. In a 250 mL Hull cell like the one used in this work, the current density distribution on the cathode surface cell can be calculated by Equation (1) [35]:

$$i = I\,(5.10 - 5.24 \log dx) \qquad (1)$$

where i is the current density (A/dm^2) at point x, I is the current passing in the cell (A), and dx is the distance (cm) between point x and the edge of the cathode that is closer to the anode.

Figure 3 shows the Ni plates resulting from the tests in the Hull cell using the baths of Table 1. The range of current densities where a black deposit was formed is presented in Table 2.

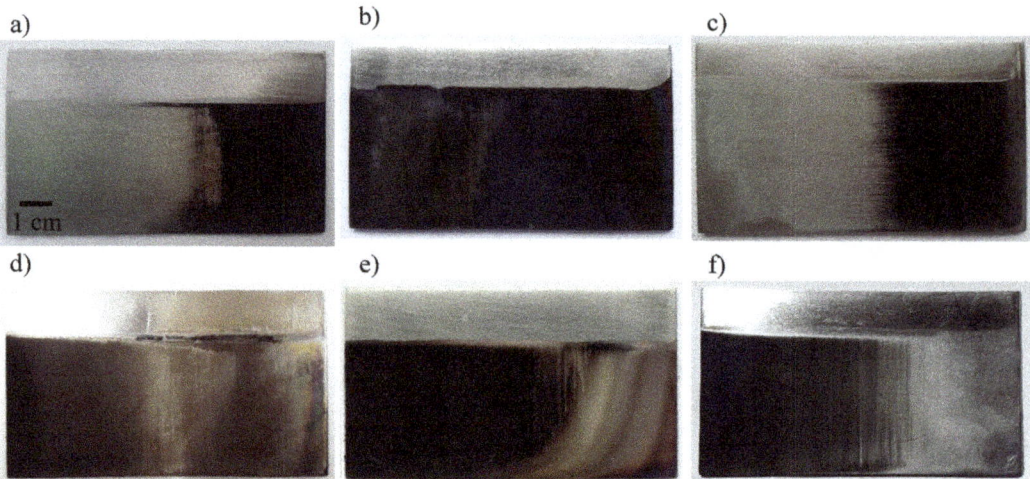

Figure 3. Hull cell samples for baths from Table 1: (**a**) bath 1, (**b**) bath 2, (**c**) bath 3, (**d**) bath 4, (**e**) bath 5, (**f**) bath 6. The higher current density was applied on to the left side of each plate.

Table 2. Hull cell results.

System	Applied Current (A)	Intervals of Current Density for Black Coatings (A/dm^2)
1	0.5	0.01–0.34
2	1	0.37–0.84
3	1	0.37–0.67
4	1	4.2–20.8
5	2	10.2–41.6
6	1	3.5–20.8

In previous works, bath 1 was used for depositing black coatings on AISI 316L stainless steel [1], and on aluminum and copper [2]. In those works, a thin layer of Ni was deposited before the black coating, which means that the black deposit was, in fact, applied to a Ni surface. This can explain the successful deposition observed in Figure 2b. Previously, the black deposit was formed with current densities ranging from 0.07 A/dm^2 to 0.26 A/dm^2, sometimes in two current steps [1,2]. Here, the electrodeposition in the Hull cell was done in just one step and the black deposit appeared in the range 0.01–0.34 A/dm^2 (Table 2). The coating presented a black matte appearance, but it was completely removed by the adhesive tape during the cross-cut test.

Regarding bath 2, in the first set of experiments, the application of the recommended 0.15 A/dm^2 (Table 1) [4] did not produce the black color. In the Hull cell, a black deposit appeared for current densities between 0.37 and 0.84 A/dm^2 (Table 2). However, it was powdery and not adherent. In bath 3, no deposition was observed in the first set of experiments. In the original work, the substrate was steel, which may explain the result. In the Hull cell, an adherent and non-uniform blackish deposit occurred between 0.37 A/dm^2 and 0.67 A/dm^2. With bath 4, a dark deposit appeared in the range between 4.2 and

20.8 A/dm^2 (Table 2). The deposit was brown-reddish, loose, and powdery. Bath 5 produced a black deposit, but high current densities were needed, between 10.2 A/dm^2 and 41.6 A/dm^2, very distant from literature value, 0.5 A/dm^2 [7]. The deposit was powdery and loose. The color was brown-reddish for lower current densities. Bath 6 showed results similar to bath 5, that is, a black deposit was formed at current densities in the interval 3.5–20.1 A/dm^2, much higher than the 0.5 A/dm^2 referred to in literature [8] and it was also powdery and non-adherent. The black coatings in references [7,8] were originally applied to molybdenum. This might explain the differences with respect to the present work. The results confirm the need to modify the original procedures considering the substrate, the particularities of the deposition cell, and the properties expected for the applied coating.

3.3. Polarization Curves

To get more insights into the electrodeposition process, cathodic polarization curves were measured on pure nickel electrodes immersed in each of the baths. All electrochemical measurements were carried out in triplicate to evaluate the reproducibility of the tests. The curves are presented in Figure 4 and the samples after the tests are shown in Figure 5. The surfaces from baths 1, 4, and 6 (Figure 5a, Figure 5d, and Figure 5f, respectively) presented black deposits while the others (Figure 5b, Figure 5c, and Figure 5e, respectively) were gray or dark gray. Only bath 1 (Figure 5a), however, produced a compact and adherent black coating. The samples from baths 4 and 6 (Figures 5d and 5f, respectively) were powdery. After the powder was removed the surface was still blackish.

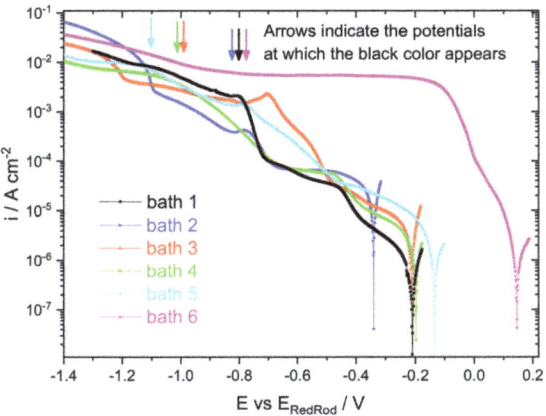

Figure 4. Polarization curves of Ni in the bath solutions presented in Table 1.

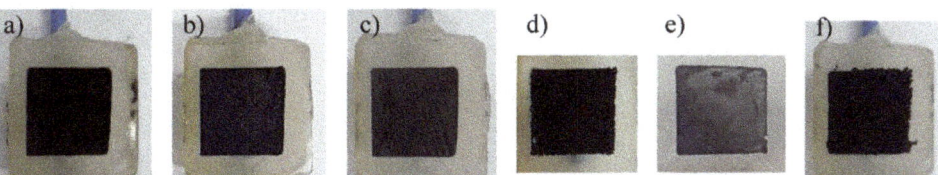

Figure 5. Ni samples (1 × 1 cm^2, embedded in insulating epoxy matrix) after polarization tests in: (**a**) bath 1, (**b**) bath 2, (**c**) bath 3, (**d**) bath 4, (**e**) bath 5, (**f**) bath 6.

The curves help identify the reactions occurring at the cathode during the electrodeposition. The potentials at which the nickel surface changed color were visually perceived and are indicated with arrows in Figure 4 and the values are presented in Table 3.

Table 3. Potential at which the surface became dark.

Solution	Potential (V_{RedRod})
1	−0.8
2	−0.8
3	−1.1
4	−1.0
5	−1.1
6	−0.8

The open-circuit potential of nickel in baths 1, 3, and 4 was -0.2 V_{RedRod}, slightly more negative in bath 2, and slightly more positive in bath 5. The potential in bath 6 was far more positive, probably due to the higher amount of cupric and persulfate ions. In this bath, upon the starting of the cathodic sweep, the reduction current increased rapidly and attained a limiting plateau around 10^{-2} A/cm^2, attributed to the reduction of Cu^{2+}(aq). Near -0.8 V_{RedRod} the current increased again due to the deposition of Ni and is where the dark color was formed. The curves measured in baths 1, 2, and 4, showed a current plateau of 40–60 µA cm^{-2} in the potential range between -0.4 V_{RedRod} and -0.6 V_{RedRod} due to the reduction of dissolved oxygen. Then, for potentials more negative than -0.7 V_{RedRod}, the current increased and a new plateau was attained at -0.8 V_{RedRod}. This increase was attributed to the deposition of metallic Ni. In baths 3 and 5, the increase in current started at less negative potentials, around -0.45 V_{RedRod}. The oxygen reduction still took place, but its current was too small compared to that of the metal deposition. A final increase in current at potentials more negative than -1.1 V_{RedRod} was due to the hydrogen reduction reaction, with H_2(g) evolution.

3.4. Optimization of the Deposition Procedure

In this work, most of the procedures indicated in Table 1 were unable to produce a compact, adherent, black coating. The changes in substrate and cell geometry require modifications in the deposition parameters to obtain the same black coatings reported in the original works. This optimization step could be applied to any of the procedures in Table 1 but it was decided to proceed only with procedure 1 because it produced the best coating in the previous sets of experiments, apart from having a simple and cheap bath composition and requiring low current density. The main problem detected with procedure 1 was the lack of adhesion during the cross-cut test in which the film was totally detached from the surface. A possible reason is that the film was deposited on the passive layer of the substrate and not directly on its pre-existent Ni player. To solve the problem, the samples were immersed in 18.5 vol% HCl for 15 min immediately before the electrodeposition to etch the surface and activate it. The bath composition was as in Table 1 (75 g/L $NiCl_2·6H_2O$ + 30 g/L NaCl) and the electrodeposition took place in just one step, with a current density of 0.143 A dm^{-2}, applied during 5 min at room temperature (~23 °C). The anodic to cathodic area ratio was A_{an}:A_{cat} = 1/1.5. The result is shown in Figure 6. The electrodeposit had a minor pulverulence on the top which, once removed with soft tissue, left a black adherent and compact film.

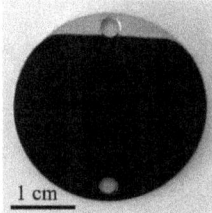

Figure 6. Nickel-plated brass sample with black nickel coating deposited from the optimized procedure 1.

3.5. SEM Images

The morphology of the coating obtained by SEM is presented in Figure 7, with the substrate before deposition (Figure 7a) and the black electrodeposit with the optimized procedure at different magnifications (Figure 7b–d). It is evident that the coating follows the substrate grain morphology and this is an indication of its thinness. A sub-micron to nanosized structure with nanoflakes is easily observed (Figure 7d). The optical interference by the surface roughness can be advanced as the reason for the black color of the coating [6,30,38].

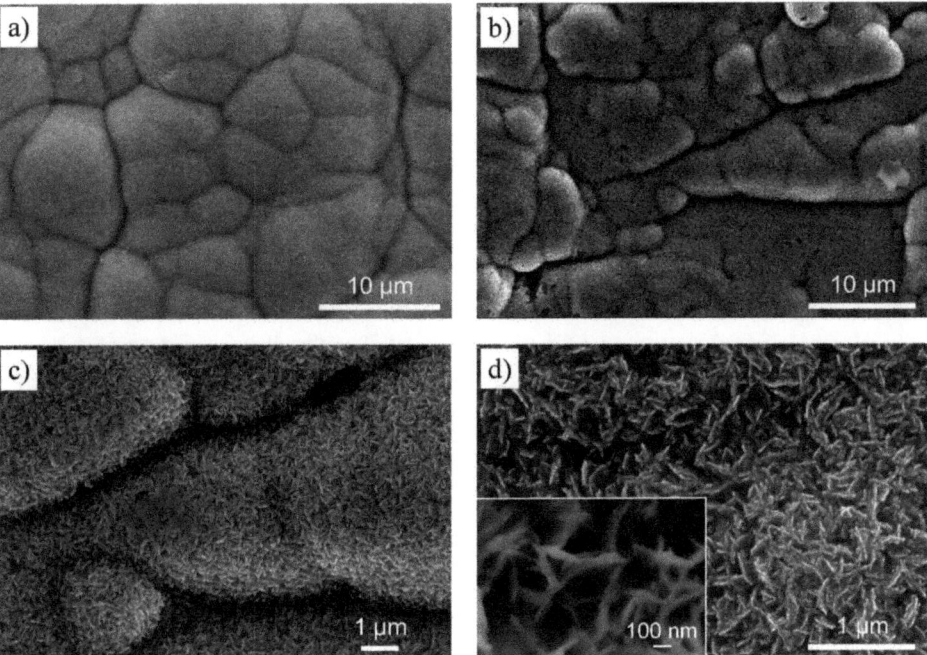

Figure 7. SEM images of the (**a**) nickel surface before deposition, (**b–d**) after deposition with the black coating at different magnifications, (**d**) nanostructured surface responsible for the black color.

3.6. Corrosion Testing in 0.5 M NaCl

The corrosion resistance of the new coating was tested in 0.5 M NaCl and compared with the Ni bare substrate. All electrochemical measurements were carried out in triplicate to evaluate the reproducibility of the tests. Figure 8a shows the polarization curves measured after 2 h of immersion. The shapes of both curves are not very different, with similar corrosion potential, E_{corr}, and a quasi-passive region with a high slope, β_a, which means that the metal oxidation will take place with difficulty as the potential increases. The corrosion current rates, i_{corr}, were obtained by the Tafel extrapolation method (using the linear region between −0.125 and 0 V for the Ni sample and −0.025 and 0.1 V for the black Ni sample) and are presented in Table 4. The curves are shifted in the current coordinates, and i_{corr} is higher in the black coating but this may be due to the high roughness of the surface, which makes the real area higher than the geometric area (the one corresponding to the 2 cm^2 of the working electrode). Admitting that the surface material is the same (nickel) then the reaction rate is the same and the variations in the measured current are just a consequence of the different areas. Based on this assumption, the ratio of i_{corr} on the coated and bare samples means that the geometric area of the black surface is around 11 times larger than the bare surface.

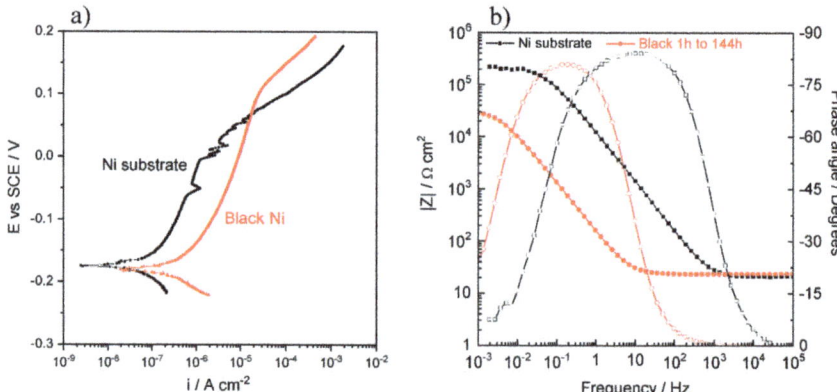

Figure 8. (a) Polarization curves measured after 2 h of immersion in 0.5 M NaCl; (b) Bode diagrams of the impedance measured after 4 h of immersion in 0.5 M NaCl. Spectra measured on the sample with black coating after 1 h, 120 h, and 144 h of immersion were identical to the spectrum after 4 h shown here.

Table 4. Parameters obtained from the polarization curves in Figure 8a.

Sample	E_{corr} (V_{SCE})	β_a (mV dec^{-1})	i_{corr} (A cm^{-2})	i_{corr} (μm year^{-1})
Ni substrate	−0.174	180.2	1.3×10^{-7}	1.4
Black Ni	−-0.182	212.3	1.5×10^{-6}	16

Considering uniform dissolution, the i_{corr} values determined by Tafel extrapolation can be converted to penetration rate (PR), using Equation (2)

$$PR = \frac{i_{corr} M_{Ni}}{nF d_{Ni}} \cdot \left(3.15 \times 10^{11}\right) \quad (2)$$

where i_{corr} is the corrosion current density (A cm^{-2}), M_{Ni} is the Ni molar weight (58.693 g mol^{-1}), n is the number of electrons involved in the oxidation reaction (2 in this case), F is the Faraday constant (96,485 C mol^{-1}), d_{Ni} is the Ni density (g cm^{-3}) of the metal, and 3.15×10^{11} is the conversion factor of year in seconds and cm in μm. The values are presented in Table 4 and show small corrosion rates, higher for the black coating probably due to the area effect. These corrosion rates are just indicative because two assumptions were made in their determination: uniform corrosion and Tafel extrapolation technique valid for these results.

The impedance response of both samples is presented in Figure 8b with Bode plots of EIS the spectra measured after 4 h in 0.5 M NaCl. It is important to refer that the spectra of the black coating measured after 1 h, 120 h, and 144 h did not change with respect to the one shown in Figure 8b, which demonstrates its stability.

In these experiments, the Ni substrate presented the highest impedance, with the response of the solution resistance at high frequencies down to 3 kHz, a capacitive response of either a passive film or the double layer capacitance, between 3 kHz and 20 mHz and finally, for low frequencies, a resistive response of about 200 kΩ cm^2, attributed to the charge transfer resistance. The black nickel sample showed a similar spectrum but with smaller charge transfer resistance and higher capacitance. These differences can be explained by the higher surface area of the black coating (not considered in the data treatment which uses the geometric area of the working electrode).

The spectra were fitted with the ZView program from Scribner Associates (USA) and the fitting parameters are presented in Table 5. The electric circuit comprised a simple $R_s(C_{dl}R_{ct})$ circuit where R_s is the solution resistance, C_{dl} is the double layer capacitance and R_{ct} is the charge transfer resistance. Constant phase elements (CPE) were used instead of capacitances, due to the non-ideal capacitive behavior of the surface [39]. The fitted parameters confirm the qualitative description presented above. In the black nickel sample the R_{ct} was 6 times lower and $Y_{0,dl}$ was 5.5 times higher than the same values found in the Ni sample. If the real area of the black nickel sample were taken into account, the resistance would increase, and the capacitance would decrease to values close to the Ni substrate. Then, their corrosion resistances would be similar.

Table 5. Parameters obtained from the impedance spectra in Figure 8b.

Sample	R_s (Ω cm^2)	$Y_{0,dl}$ (Ω^{-1} s^n cm^{-2})	n_{dl}	R_{ct} (Ω cm^2)	10^4 χ^2
Ni substrate	21.0	1.51×10^{-5}	0.937	2.08×10^5	11
Black Ni	22.4	8.34×10^{-5}	0.933	3.32×10^4	3.8

4. Conclusions

In this work, six different electrodeposition procedures for producing nickel black coatings were compared. It was verified that the direct application of the conditions reported in the literature either did not produce any deposit or produced pulverulent and loose deposits. In one case a compact black coating was deposited but the adhesion to the substrate was poor. It became evident that the results could be due to the change in substrate and cell geometry with respect to the original works. Hull cell experiments identified ranges where all procedures were able to produce black deposits. Still, it was clear that optimization was needed to produce a coating with the desired finish quality regarding color, compactness, and adhesion. Work progressed with the procedure that gave the best results, and the final optimal conditions were found with a bath consisting of 75 g/L $NiCl_2 \cdot 6H_2O$ + 30 g/L NaCl, a current density of 0.143 A/dm^2 for 5 min of deposition at room temperature. Immediately before the electrodeposition, an acid pretreatment was necessary to warrant adhesion to the substrate. The black surface showed microroughness with nanoflakes. The black color of the coating was attributed to the optical interference caused by the surface texture. Polarization curves and electrochemical impedance spectroscopy showed good corrosion resistance. Work continues to tune the conditions for coating gloss as particular values are important for the different applications.

Author Contributions: Conceptualization and methodology: M.M.M., A.C.B., and M.G.S.F.; Investigation: M.M.M.; Writing—original draft preparation, review, and editing: M.M.M., R.M., A.C.B., A.M., P.O., J.T., and M.G.S.F.; Supervision: A.C.B. and M.G.S.F. All authors have read and agreed to the published version of the manuscript.

Funding: This research was funded by project ON-SURF—Mobilizar Competências Tecnológicas em Engenharia de Superfícies, Projeto n.º POCI-01-0247-FEDER-024521.

Institutional Review Board Statement: Not applicable.

Informed Consent Statement: Not applicable.

Data Availability Statement: The data used to support the findings of this study are available from the correspondence authors upon request.

Acknowledgments: This work was developed within the scope of the project CICECO-Aveiro Institute of Materials, UIDB/50011/2020 and UIDP/50011/2020, financed by national funds through the Foundation for Science and Technology/MCTES. A.C.B. acknowledges FCT—Fundação para a Ciência e a Tecnologia, I.P., in the scope of the framework contract foreseen in the numbers 4, 5, and 6 of the article 23, of the Decree-Law 57/2016, of 29 August, changed by Law 57/2017, of 19 July. M.M.M. acknowledges FCT—Fundação para a Ciência e a Tecnologia for the scholarship (BPD-UA-A n°24521 ERDF).

Conflicts of Interest: The authors declare no conflict of interest.

References

1. Lira-Cantu, M.; Sabio, A.M.; Brustenga, A.; Gómez-Romero, P. Electrochemical deposition of black nickel solar absorber coatings on stainless steel AISI316L for thermal solar cells. *Sol. Energy Mater. Sol. Cells* **2005**, *87*, 685–694. [CrossRef]
2. Wäckelgård, E. Characterization of black nickel solar absorber coatings electroplated in a nickel chlorine aqueous solution. *Sol. Energy Mater. Sol. Cells* **1998**, *56*, 35–44. [CrossRef]
3. Wernick, S.; Pinner, R. *The Surface Treatment and Finishing of Aluminium and Its Alloys*; Robert Draper Ltd: Teddington, UK, 1972; Volume 2.
4. Dennis, J.K.; Such, T.E. *Nickel and Chromium Plating*, 2nd ed.; Butterworth&Co.: London, UK, 1986; pp. 48–50.
5. Ibrahim, M.A.M. Black nickel electrodeposition from a modified Watts bath. *J. Appl. Electrochem.* **2006**, *36*, 295. [CrossRef]
6. Karuppiah, N.; John, S.; Natarajan, S.; Sivan, V. Characterization of electrodeposited Nickel-Cobalt selective black coatings. *Bull. Electrochem.* **2002**, *18*, 295–298.
7. Aravinda, C.L.; Bera, P.; Jayaram, V.; Sharma, A.K.; Mayanna, S.M. Characterization of electrochemically deposited Cu-Ni black coatings. *Mater. Res. Bull.* **2002**, *37*, 397–405. [CrossRef]
8. Aravinda, C.L.; Mayanna, S.M.; Bera, P.; Jayaram, V.; Sharma, A.K. XPS and XAES investigations of electrochemically deposited cu-Ni solar selected black coatings on molybdenum substrate. *J. Mater. Sci. Lett.* **2002**, *21*, 205–208. [CrossRef]
9. Takadoum, J. Black coatings: A review. *Eur. Phys. J. Appl. Phys.* **2010**, *52*, 30401. [CrossRef]
10. Lelevic, A.; Walsh, F.C. Electrodeposition of Ni-P composite coatings: A review. *Surf. Coat. Technol.* **2019**, *378*, 124803. [CrossRef]
11. Gogna, P.K.; Chopra, K.L. Structure-dependent thermal and optical properties of black nickel coatings. *Thin Solid Film.* **1979**, *57*, 299–302. [CrossRef]
12. Lee, T.K.; Kim, D.H.; Auh, P.C. The optical characteristics of black chrome solar selective films coated by the pulse current electrolysis method. *Sol. Energy Mater. Sol. Cells* **1993**, *29*, 149–161. [CrossRef]
13. Popov, B.N.; White, R.E.; Slavkov, D.; Koneska, Z. Reduction of chromium (VI) when solar selective black chromium is deposited in the presence of organic additive. *J. Electrochem. Soc.* **1992**, *139*, 91. [CrossRef]
14. Gigandet, M.P.; Faucheu, J.; Tachez, M. Formation of black chromate conversion coatings on pure and zinc alloy electrolytic deposits: Role of the main constituents. *Surf. Coat. Technol.* **1997**, *89*, 285–291. [CrossRef]
15. Dubpernell, G. *Electrodeposition of Chromium from Chromic Acid Solutions*; Pergamon Press Inc.: New York, NY, USA, 1977. [CrossRef]
16. Kimbrough, D.E.; Cohen, Y.; Winer, A.M.; Creelman, L.; Mabuni, C. A Critical assessment of chromium in the environment. *Crit. Rev. Environ. Sci. Technol.* **1999**, *29*, 1–46. [CrossRef]
17. Pellerin, C.; Booker, S.M. Reflections on hexavalent chromium: Health hazards of an industrial heavyweight. *Environ. Health Perspect.* **2000**, *108*. [CrossRef] [PubMed]
18. Rahman, Z.; Singh, V.P. The relative impact of toxic heavy metals (THMs) (arsenic (As), cadmium (Cd), chromium (Cr)(VI), mercury (Hg), and lead (Pb)) on the total environment: An overview. *Environ. Monit. Assess.* **2019**, *191*, 419. [CrossRef]
19. Vaiopoulou, E.; Gikas, P. Regulations for chromium emissions to the aquatic environment in Europe and elsewhere. *Chemosphere* **2020**, *254*, 126876. [CrossRef]
20. Kalidhasan, S.; Kumar, A.S.K.; Rajesh, V.; Rajesh, N. The journey traversed in the remediation of hexavalent chromium and the road ahead toward greener alternatives—A perspective. *Coord. Chem. Rev.* **2016**, *317*, 157–166. [CrossRef]
21. Gharbi, O.; Thomas, S.; Smith, C.; Birbilis, N. Chromate replacement: What does the future hold? *Mater. Degrad.* **2018**, *2*, 12. [CrossRef]
22. Wilbur, S.; Abadin, H.; Fay, M.; Yu, D.; Tencza, B.; Ingerman, L.; Klotzbach, J.; James, S. *Toxicological Profile for Chromium*; Agency for Toxic Substances and Disease Registry (US): Atlanta, GA, USA, 2012.
23. World Health Organization. *Chromium—Environmental Health Criteria n° 61*; WHO: Geneva, Switzerland, 1988.
24. World Health Organization. Chapter 6.4—Chromium. In *Air Quality Guidelines for Europe*, 2nd ed.; European Series No. 91; WHO Regional Publications: Copenhagen, Denmark, 2001.
25. Von Burg, R.; Liu, D. Chromium and hexavalent chromium. *J. Appl. Toxicol.* **1993**, *13*, 225–230. [CrossRef]
26. Nunes, R.A.X.; Costa, V.C.; Sade, W.; Araújo, F.R.; Silva, G.M. Selective surfaces of black chromium for use in solar absorbers. *Mater. Res.* **2018**, *21*, e20170556. [CrossRef]
27. Patel, S.N.; Inal, O.T. Optimization and microstructural analysis of black-zinc-coated aluminum solar collector coatings. *Thin Solid Film.* **1984**, *113*, 47–57. [CrossRef]
28. Li, G.-Y.; Lian, J.-S.; Niu, L.-Y.; Jiang, Z.-H. A zinc and manganese phosphate coating on automobile iron castings. *ISIJ Int.* **2005**, *45*, 1326–1330. [CrossRef]
29. Jahan, F.; Islam, M.H.; Smith, B.E. Band gap and refractive index determination of Mo-black coatings using several techniques. *Sol. Energy Mater. Sol. Cells* **1995**, *37*, 283–293. [CrossRef]
30. Lizama-Tzec, F.I.; Macías, J.D.; Estrella-Gutiérrez, M.A.; Cahue-López, A.C.; Arés, O.; de Coss, R.; Alvarado-Gil, J.J.; Oskam, G. Electrodeposition and characterization of nanostructured black nickel selective absorber coatings for solar–thermal energy conversion. *J. Mater. Sci. Mater. Electron.* **2015**, *26*, 5553–5561. [CrossRef]
31. Li, X.J.; Cai, C.; Song, L.; Li, J.; Zhang, Z.; Xue, M.; Liu, Y. Electrodeposition and characterization of nano-structured black nickel thin films. *Trans. Nonferrous Met. Soc. China* **2013**, *23*, 2300–2306. [CrossRef]
32. Delaunois, F.; Vitry, V.; Bonin, L. (Eds.) *Electroless Nickel Plating: Fundamentals to Applications*, 1st ed.; CRC Press: Boca Raton, FL, USA, 2019. [CrossRef]

33. Chappé, J.M.; Vaz, F.; Cunha, L.; Moura, C.; Marco de Lucas, M.C.; Imhoff, L.; Bourgeois, S.; Pierson, J.F. Development of dark Ti(C,O,N) coatings prepared by reactive sputtering. *Surf. Coat. Technol.* **2008**, *203*, 804–807. [CrossRef]
34. Andritschky, M.; Atfeh, M.; Pischow, K. Multilayered decorative a-C:H/CrC coating on stainless steel. *Surf. Coat. Technol.* **2009**, *203*, 952–956. [CrossRef]
35. Deutschen Normen. *Galvanisierungsprüfung mit der Hull-Zelle, DIN50957*; Deutsches Institut für Normung: Berlin, Germany, 1978.
36. Hull, R.O. Current density range characteristics-Their determination and application. *Proc. Am. Electroplat. Soc.* **1938**, *27*, 52.
37. I.O. for Standardization. Cross-cut test ISO 2409:2007. *Br. Stand. Inst.* **2013**, *3*, 9–11.
38. Estrella-Gutiérrez, M.A.; Lizama-Tzec, F.I.; Arés-Muzio, O.; Oskam, G. Influence of a metallic nickel interlayer on the performance of solar absorber coatings based on black nickel electrodeposited onto copper. *Electrochim. Acta* **2016**, *213*, 460–468. [CrossRef]
39. Orazem, M.E.; Tribollet, B. *Electrochemical Impedance Specstroscopy*, 2nd ed.; John Wiley & Sons: Hoboken, NJ, USA, 2017; p. 395.

Article

Synthesis of Silver Nanoparticles with Gemini Surfactants as Efficient Capping and Stabilizing Agents

Bogumił Brycki *, Adrianna Szulc and Mariia Babkova

Department of Bioactive Products, Faculty of Chemistry, Adam Mickiewicz University Poznan, 61-614 Poznan, Poland; adaszulc@amu.edu.pl (A.S.); mariiababkova97@gmail.com (M.B.)
* Correspondence: brycki@amu.edu.pl; Tel.: +48-61-829-1694

Abstract: The scientific community has paid special attention to silver nanoparticles (AgNPs) in recent years due to their huge technological capacities, particularly in biomedical applications, such as antimicrobials, drug-delivery carriers, device coatings, imaging probes, diagnostic, and optoelectronic platforms. The most popular method of obtaining silver nanoparticles as a colloidal dispersion in aqueous solution is chemical reduction. The choice of the capping agent is particularly important in order to obtain the desired size distribution, shape, and dispersion rate of AgNPs. Gemini alkylammonium salts are named as multifunctional surfactants, and possess a wide variety of applications, which include their use as capping agents for metal nanoparticles synthesis. Because of the high antimicrobial activity of gemini surfactants, AgNPs stabilized by this kind of surfactant may possess unique and strengthened biocidal properties. The present paper presents the synthesis of AgNPs stabilized by gemini surfactants with hexadecyl substituent and variable structure of spacer, obtained via ecofriendly synthesis. UV-Vis spectroscopy and dynamic light scattering were used as analyzing tools in order to confirm physicochemical characterization of the AgNPs (characteristic UV-Vis bands, hydrodynamic diameter of NPs, polydispersity index (PDI)).

Keywords: silver nanoparticles; gemini surfactant; green synthesis

1. Introduction

Over the past twenty years, silver nanoparticles (AgNPs) have become increasingly popular due to their special physical, chemical, and biological properties. It is characteristic that, depending on the surface to volume ratio, the properties of nanosilver change, and they have been exploited for different purposes [1,2]. AgNPs offer promising prospects for a wide range of applications, such as antimicrobials against bacteria [3–12], fungi [10,13–17], and viruses [9,18–23]. Silver nanoparticles can also be helpful in managing the ongoing pandemic of COVID-19 (Coronavrus Disease 2019) caused by the SARS-CoV-2 (severe acute respiratory syndrome coronavirus 2) [24,25]. AgNPs have been tested as biomedical device coatings [12,26–28], combating multidrug-resistant cancer [12,29], drug-delivery carriers [12,30], and imaging probes in ultrasensitive analysis [26,31–34]. They may also find applications in sensing [27,31,35,36] and chemical catalysis [32,35,37–39]. Due to the great interest of AgNPs, different synthetic methods have been developed, such as physical, chemical, and biological approaches [26,40,41]. Among them, chemical reduction is the most frequently used synthesis method of silver nanoparticles, with high productivity and low costs [39,41]. In this approach, stable colloids in water or non-aqueous solvents can be obtained. Generally, synthesis of AgNPs in solution requires three particular components: metal precursors (silver nitrate, perchlorate, chloride, trifluoroacetate or $(PPh_3)_3AgNO_3$), reducers, and stabilizers [42]. Several reductants are commonly used for the preparation of the AgNPs: sodium borohydride, sodium citrate, glucose, N,N-dimethylformamide, hydrazine, polyols, (e.g., ethylene glycol, diethylene glycol), formaldehyde [42]. One important factor is to apply capping agents to stabilize dispersive NPs, thus protect them against their agglomeration [41]. In general, the shape and size distribution of the synthesized nanoparticles are controlled by altering the method and conditions of the synthesis, reducing and stabilizing agents, their concentrations, and molar ratios [43]. Recently, different types of surfactants started to be used as stabilizing agents for AgNPs. We should mention that a proper stabilizer and/or reaction time is crucial for the obtaining of stable and small size Ag nanoparticles [44].

Dimeric quaternary alkylammonium salts, named by Menger and Littau as gemini, are modern type of surfactants [45]. These compounds consist of two long alkyl substituents and two cationic headgroups connected by a linker. The connector may have various structures: short, long, flexible, stiff, polar (e.g., functionalized by heteroatoms), and nonpolar (e.g., hydrocarbon chains or rings) [46–50]. Spacer and hydrophobic chain are very important units in gemini surfactant architecture. They are a keystone in the adsorption of dimeric surfactants on surfaces and interfaces, in creation of micelles of different shapes [48,49,51–54]. Considering the structure of gemini surfactants, they exhibit a higher efficacy in contrast to monomeric (conventional) surfactants, such as decreased critical micelle concentration (CMC), greater ability to reduce surface tension, or promote antimicrobial activity [48,50,55–57]. From an ecological point of view, dimeric alkylammonium salts are considerably less toxic to marine organisms than quaternary ammonium salts [58,59], and can be biodegradable [48,60–62]. Considering the above properties, and the fact that the use of gemini surfactant provides the desired effect at a lower concentration in comparison to monomeric analogues, the use of dimeric alkylammonium salts is correlated with the greenolution idea in chemistry [63,64]. Gemini surfactants, due to their antimicrobial properties against bacteria [48,55,56,65–70] and fungi [47,48,56,66,71], can be used as microbicides or biofilm eradication agents in many areas of life and industries [66,72–74].

In reference to the growing interest in silver nanoparticles and gemini surfactants, we focused on the different procedures for obtaining silver nanoparticles capped with gemini surfactants, having a different structure in spacer unit, obtained in easy, low-cost, and green synthesis. Detailed spectroscopic analysis and surface activity of the gemini surfactants, as well as analysis of AgNPs based on a UV-Vis, and dynamic light scattering (DLS) study were presented.

2. Materials and Methods

2.1. Material Used

N,N-Dimethyl-N-hexadecylamine, bis(2-bromoethyl)ether, 1,6-dibromohexane, silver nitrate, and sodium borohydride, were obtained from Sigma-Aldrich (Poznan, Poland). Acetonitrile and diphosphorus pentoxide were purchased from VWR Chemicals (Gdansk, Poland). Reagent purity was at least 95%. All materials were used without prior purification.

2.2. Synthesis

2.2.1. Synthesis of Gemini Surfactants

The 16-6-16 (1,6-hexamethylene-bis(N-hexadecyl-N,N-dimethylammonium bromide)): the N,N-dimethyl-N-hexadecylamine (10 g; 37 mmol) and 1,6-dibromohexane (4.5 g; 18.8 mmol) were placed in a 250 mL round-bottom flask. Reaction mixture was stirred without the solvent at room temperature for 2 h. Then, the reaction mixture was left to solidify completely. The crude product was purified by crystallization from acetonitrile, filtered, and dried in a vacuum desiccator over P_2O_5 and in an oven at 60 °C. Reaction yield was over 98%.

The 16-O-16 (3-oxa-1,5-pentane-bis(N-hexadecyl-N,N-dimethylammonium bromide)): the N,N-dimethyl-N-hexadecylamine (10 g; 37 mmol) and bis(2-bromoethyl)ether (4.36 g; 18.5 mmol) were placed in a 250 mL round-bottom flask. Reaction mixture was stirred without the solvent at room temperature for 2 h. Then, the reaction mixture was left to solidify completely. The crude product was purified by crystallization from acetonitrile, filtered, and dried in a vacuum desiccator over P_2O_5 and in an oven at 60 °C. Reaction yield was over 95%.

2.2.2. Synthesis of Silver Nanoparticles

Experiments 1–6: 40 mL of 0.75 mM silver nitrate was mixed with 10 mL of specific amount of gemini surfactant (16-6-16 or 16-O-16) corresponding to three different molar ratio nAg/nGemini = 2.5; 5; 10. The mixture was stirred for 5 min, and then 5 mL of 1 mM sodium borohydride cold solution was added dropwise. Color changed to intense orange. After adding all of the solution, stirring occurred for another 20 min.

Experiment 7: 15 mL of 4 mM $NaBH_4$ and 10 mL of 0.2 mM 16-6-16 solutions were placed in a 100 mL Erlenmeyer flask and stirred for 5 min. Then 5 mL of 2 mM $AgNO_3$ was added dropwise to the flask. Stirring continued for another 20 min.

Experiment 8: 5 mL of 12 mM $NaBH_4$ was placed in a 100 mL Erlenmeyer flask with 20 mL of water. Solution of 10 mL of 1 mM $AgNO_3$ and 10 mL of 0.2 mM 16-6-16 was prepared separately and added dropwise into $NaBH_4$. After adding all of the solution, stirring occurred for another 20 min.

All of the experiments were repeated three times in order to confirm the reproducibility of the synthetic procedures. The AgNPs stabilized by gemini surfactants were stable for at least three months, during which no changes in the UV-vis and DLS were observed.

2.3. Experimental Methods

2.3.1. Analysis of Dimeric Alkylammonium Salts

The melting point of the synthesized gemini surfactants was measured on Stuart SMP30 apparatus (Staffordshire, UK), using capillary with one sealed side. The measurements give the values with resolution of 1.0 °C. The elemental analysis measurements were performed on a FLASH 2000 elemental analyzer (Delft, The Netherlands) with a thermal conductivity detector. The nuclear magnetic resonance (NMR) spectra were measured with a Varian VNMR-S 400 MHz (Oxford, UK), operating at 403 and 101 MHz for ^1H and ^{13}C respectively, attendant with software Vnmr VERSION 2.3 REVISION A (Varian, Oxford, UK). The chemical shifts were performed in $CDCl_3$ relative to an internal standard—TMS (tetramethylsilane). The Fourier Transform Infrared Spectroscopy (FTIR) spectra were performed on FT-IR Bruker IFS 66v/S (Poznan, Poland) apparatus. All of the samples were tested in solid state in the form of tablets with potassium bromide.

The critical micelle concentration values were determined by conductivity analysis using a Conductivity Meter Elmetron CC-505 (Zabrze, Poland). The apparatus was calibrated by a standard (147 µS/cm in 298.15 K). All samples were obtained using deionized water. Conductivity measurements were carried out at 25 °C. The titration was repeated at least three times for each compound.

2.3.2. Analysis of Silver Nanoparticles

UV-Vis absorption spectra of all the AgNPs solutions capped with dimeric ammonium salts were recorded on Varian Cary 50 Scan UV-Vis spectrophotometer (Varian, Oxford, UK) at 25 °C, operated with Cary WinUV software (Varian, Oxford, UK) with quartz cuvettes of 1 cm path length. In all cases, samples were diluted 5-fold with deionized water, to decrease the absorbance to the range suitable for UV-Vis measurements.

Dynamic light scattering measurements and polydispersity index (PDI) were performed at 25 °C using Zetasizer, Nano-ZS Malvern Instruments (Malvern, UK) with a He-Ne laser (633 nm, 4 mW) equipped with a built-in termo-controller. Ten repeated measurements were conducted for each sample.

3. Results and Discussion

3.1. Synthesis and Spectroscopic Characterization of Gemini Surfactants

Gemini surfactants tested in this paper as capping agents for AgNPs were received by alkylation of *N,N*-dimethyl-*N*-hexadecylamine with 1,6-dibromohexane or bis(2-bromoethyl) ether, for 16-6-16, and 16-O-16 respectively (Figure 1). A synthetic procedure developed in our laboratory assumes carrying out the reaction without a solvent at room temperature [50,75]. These reactions proceed according to the nucleophilic substitution mechanism S_N2. The rate of these reactions depends on the substrates concentrations. In case of the reaction without a solvent, the concentration of the reactants is the highest possible. Such reactions take place with high efficiency without waste. These reactions are in accordance with the greenolution approach, because they reduce the use of solvent to a minimum and minimize costs.

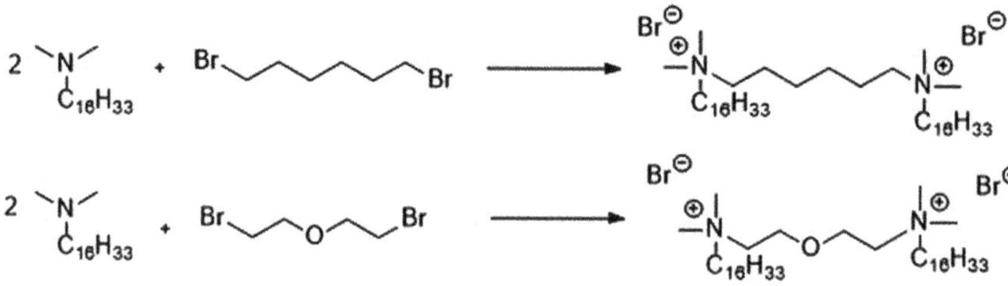

Figure 1. Synthesis of 16-6-16 and 16-O-16.

The structure and purity of gemini surfactants obtained were confirmed by ^{1}H NMR, ^{13}C NMR, FTIR, and elemental analysis.

The **16-6-16**: mp 222–223 °C; elemental analysis (%) calcd. for $C_{42}H_{90}Br_2N_2$: C 64.43, H 11.58, N 3.58; found C 64.72, H 11.68, N 3.40; ^{1}H NMR (403 MHz, CDCl$_3$) 3.73 (bs, 4H, N$^+$CH$_2$(spacer)), 3.49 (m, 4H, -N$^+$CH$_2$-), 3.39 (s, 12H, -N$^+$(CH$_3$)$_2$), 2,01 (bs, 4H, -N$^+$CH$_2$CH$_2$-(spacer)), 1.72 (bs, 4H, -N$^+$CH$_2$CH$_2$-),1.57 (bs, 4H, -N$^+$CH$_2$CH$_2$-CH$_2$(spacer)), 1.36-1.25 (m, 52H, -CH$_2$(CH$_2$)$_{13}$CH$_3$), 0.88 (t, 6H, -CH$_2$CH$_3$); ^{13}C NMR (101 MHz, CDCl$_3$): 65.9 (-N$^+$CH$_2$spacer), 64.06 (-N$^+$CH$_2$-), 50.98 (-N$^+$CH$_3$), 31.89 (-CH$_2$CH$_2$CH$_3$), 29.61, 29.59, 29.56, 29.34, 29.27, 29.45, 29.37, 29.12 (-N$^+$(CH$_2$)$_2$(CH$_2$)$_{11}$-), 26.86 (-N$^+$CH$_2$CH$_2$-), 24.37 – (N$^+$CH$_2$CH$_2$-(spacer)), 22.6 (-N$^+$CH$_2$CH$_2$-CH$_2$(spacer)), 21.61 (-CH$_2$CH$_3$), 14.11 (-CH$_2$CH$_3$).

The 16-O-16: mp 240-242 °C; elemental analysis (%) calcd. for $C_{40}H_{86}Br_2N_2O$: C 62.32, H 11.24, N 3.63; found C 61.41, H 11.68, N 3.20; ^{1}H NMR (403 MHz, CDCl$_3$) 4.36 (bs, 4H, OCH$_2$), 4.05 (bs, 4H, N$^+$CH$_2$CH$_2$O-), 3.67-3.62 (m, 4H, -N$^+$CH$_2$-), 3.46 (s, 12H, -N$^+$(CH$_3$)$_2$), 1.73 (bs, 4H, -N$^+$CH$_2$CH$_2$-), 1.47-1.10 (m, 52H, -CH$_2$(CH$_2$)$_{13}$CH$_3$), 0.88 (t, 6H, -CH$_2$CH$_3$); ^{13}C NMR (101 MHz, CDCl$_3$): 65.92(-N$^+$CH$_2$CH$_2$O-), 64.59 (-OCH$_2$-), 64.03 (-N$^+$CH$_2$-), 51.57 (-N$^+$CH$_3$), 31.82 (-CH$_2$CH$_2$CH$_3$), 29.61, 29.58, 29.56, 29.54, 29.45, 29.37, 29.25, 26.25 (-N$^+$(CH$_2$)$_2$(CH$_2$)$_{11}$-), 22.83 (-N$^+$CH$_2$CH$_2$-), 22.58 (-CH$_2$CH$_3$), 14.00 (-CH$_2$CH$_3$).

FTIR spectra for tested dimeric alkylammonium salts (Figure 2) show regular bands for the asymmetric (v_{as}) and symmetric (v_s) stretching vibrations of methyl and methylene groups at 2916–2846 cm^{-1}, as well as bands for the deformation vibration (δ) of methyl groups at 1472–1464 cm^{-1}. The 16-O-16 spectra shows usual peaks of C-O stretching vibration at 1146 cm^{-1}. Due to hydrophilicity (ability to absorb water) of gemini surfactants the strong absorption at the 3500–3350 cm^{-1} region from the -OH stretching vibrations of water molecule is detected in FTIR spectra.

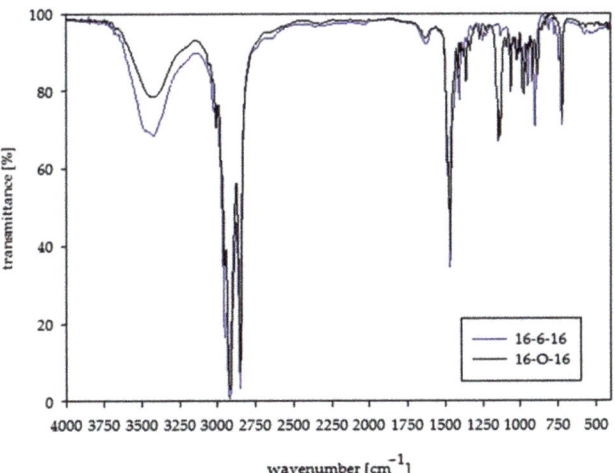

Figure 2. FTIR spectra for 16-6-16 and 16-O-16.

3.2. Surface Properties of Gemini Surfactants

The basic ability of surface active agents is their tendency to be adsorbed at interfaces. Mechanism of surface action of dimeric surfactants based on the adsorption of ammonium cations into a polar phase, and hydrocarbon chains in a nonpolar phase [48]. The keystone of all surfactants research is determination of their critical micelle concentration [49], the lowest concentration at which particles rapidly aggregate into micelles. The fact that dimeric alkylammonium salts have much lower CMC values than monomeric ones, creates great application possibilities for them [52,63,76–79]. The determination of CMC is also important, because toxicity of gemini rises when their concentration surpasses CMC [80].

The CMC was determined using a conductometric titration, creating dependency graphs of the characteristic conductivity in water of the obtained compounds as a function of the concentration. This relationship generates two lines with different slopes. The straight line with higher inclination illustrates behavior before micellization, whereas the line with a smaller slope indicates the process of micellization (Figure 3). The intersection of lines formed as a result of linear regression defines CMC [50,59].

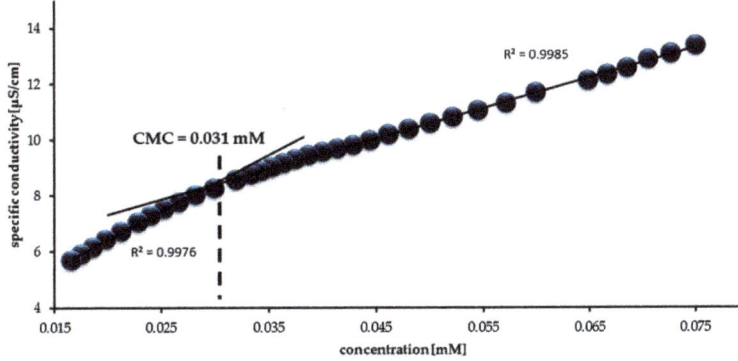

Figure 3. Specific conductivity versus surfactant concentration in water for 16-O-16.

The values for the slope ratios of linear regression in pre- and post-micellization regions allow estimating the degree of ionization (α) and the degree of counterion binding parameter (β). These parameters demonstrate the capability of the counterion binding on the micelles [50,59]. Knowing the values of CMC and β allow for calculation of Gibbs free energy of micellization ($\Delta G°_{mic}$) [81]. The experimental values of CMC, α, β and $\Delta G°_{mic}$ for the investigated gemini surfactants are given in Table 1.

Table 1. Critical micelle concentration, degree of ionization, counterion binding parameter, and Gibbs free energy of micellization in 25 °C.

Surfactant	CMC (mM)	α	β	$\Delta G°_{mic}$ (kJ/mol)
16-6-16	0.034	0.55	0.45	−50.16
16-O-16	0.031	0.40	0.60	−58.32

Critical micelle concentration of dimeric surfactants depends on many structural elements: length and structure of spacer and substituents and type of anion. It is well known that gemini surfactants with long (hexadecyl, octadecyl) substituents possess lower CMC values than analogues with shorter hydrocarbon chain [48–50,77,82–86]. CMCs of dimeric alkylammonium salts tested in this work are over thirty times lower, than CMC of 12-6-12 (1,6-hexamethylene-bis(N-hexadecyl-N,N-dimethylammonium bromide)) which is 0.98 (mM) [57]. Surface activity of our gemini surfactants, although they have different spacer structure, is comparable. This result arein good correlation to our previous result—that an element of the structure, which essentially affects the surface activity, is the length of the substituent, not the structure of the linker [50].

The 16-O-16 possesses a lower degree of ionization value than 16-6-16. The small α is caused by the stronger binding of the counterion to the aggregates, suggesting better packing of the head groups and higher surface charge density at the interface [50,51]. Both obtained gemini surfactants have negative $\Delta G°_{mic}$, indicating that the micellization process is spontaneous [50,55]. Surfactant with oxygen-functionalized spacer (16-O-16) shows greater tendency to form micelles, because in this case $\Delta G°_{mic}$ is lower.

3.3. Preparation and Characterization of Silver Nanoparticles

AgNPs were prepared by chemical synthetic procedure under different conditions. We used: silver salt (AgNO$_3$) as a metal precursor, sodium borohydride (NaBH$_4$) as a reductant and gemini surfactants (16-6-16 in experiments 1–3 and 7–8, 16-O-16 in experiments 4–6) as stabilizing agents (Figure 4). Experiments 1–6 were performed according to preparation described by Pisárčik [87]. Different molar ratios of AgNO$_3$ and gemini surfactants were applied to study the formation of AgNPs (Table 2). In these cases, nAgNO$_3$ was taken in excess of 6-fold to nNaBH$_4$. Experiments 7 and 8 were carried out by another procedure, with 16-6-16 as stabilizing agent in silver-gemini surfactants molar ratio 5 (Table 2). In these cases, nNaBH$_4$ was taken in excess of 6-fold to nAgNO$_3$.

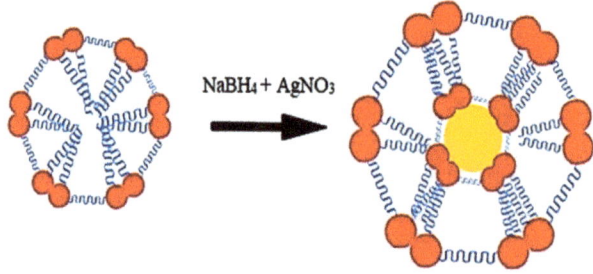

Figure 4. Schematic representation of the formation of layer of gemini surfactants, surrounding the Ag nanoparticles.

Table 2. Experimental data of different procedures for AgNPs obtaining.

Experiment	Gemini Surfactant	Molar Ratio nAg/nGemini	Concentration of $AgNO_3$ [mM]
1	16-6-16	2.5	0.75
2	16-6-16	5	0.75
3	16-6-16	10	0.75
4	16-O-16	2.5	0.75
5	16-O-16	5	0.75
6	16-O-16	10	0.75
7	16-6-16	5	2
8	16-6-16	5	1

The color of AgNPs solution depends on the degree of dilution of the colloids, starting from yellow and going to intense orange with the increase of colloids concentration. Gemini surfactant allows for keeping the nanodispersion homogeneous, with a color depending on the nanodispersion dilution. In our work, in all experiments, the intense orange color was formed, showing that the reaction was followed by the formation of nanoparticles [87–90]. The color of AgNPs solution is suggested to appear due to localized surface plasmon resonance (LSPRs). LSPR is a collective excitation of the free electrons in the conduction band around the surface of the nanoparticles. The electrons are specified to particular vibration modes, which depend on particle size and shape. Therefore, AgNPs could be detected and characterized by UV-Vis. With increasing of particle size, the absorption wavelength shifts to longer wavelengths [91].

UV-Vis spectroscopic analysis is an essential technique for observation of AgNPs formation and stability in solution. The absorbance peak of silver nanoparticles usually occurs at wavelength range of 350–450 nm, and moves to longer wavelengths with progressive particle size [91]. In Figure 5a, the UV-Vis spectra of AgNPs, which were stabilized with 16-6-16 is presented. The spectra were measured for three various Ag-to-gemini surfactant molar ratios: nAg/n16-6-16 = 2.5; 5; 10. We can see that from the highest gemini concentration (experiment 1) to the lowest (experiment 3), the peak of absorbance increases gradually: 0.286, 0.332 and 0.359. As it is known, the absorbance is directly connected with the concentration, thus, since the difference is not significant, it can be concluded that 16-6-16 has a strong stabilizing effect on AgNPs in the extensive range of nAg/n16-6-16 molar ratio values. However, the nAg/n16-6-16 molar ratio at 2.5 indicates a slightly weak stabilizing effect since the concentration of AgNPs is not high.

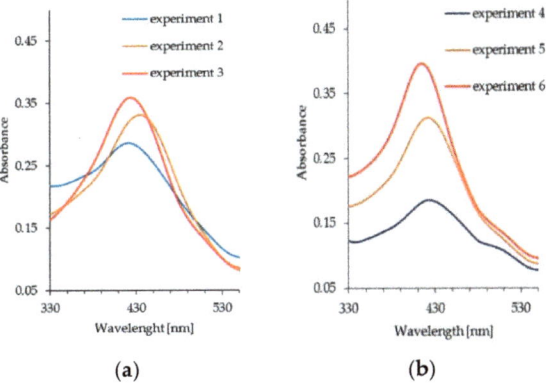

Figure 5. UV-Vis spectra of silver nanoparticles stabilized with gemini surfactants (a) 16-6-16; (b) 16-O-16 at different nAg/nGemini values.

Figure 5b shows the UV-Vis absorption spectra of nanoparticles stabilized with 16-O-16. These AgNPs were synthesized similar to the previous ones, but 16-O-16 was taken as capping agent (experiment 4–6). We observe that the peak of absorbance gradually increases with the decrease of gemini surfactant concentration as well, as in the case of experiments 1–3. The peaks of absorbance for experiments 4, 5, and 6 are 0.185, 0.313, and 0.397, respectively. The stabilizing effect of 16-O-16 on AgNPs is stronger for molar ratios of nAg/n16-O-16 = 5 and 10, whereas it has weaker effect for reaction with nAg/n16-O-16 molar ratio of 2.5. Same trend was observed for of 16-6-16.

UV–Vis absorption spectra of AgNPs obtained in experiments 7 and 8 are shown in Figure 6. Strong absorption peaks at approximately 410 and 419 nm come from the surface plasmon absorption of nanosized silver particles. These spectra stand out; the good symmetric absorption peaks with a nearly unaltered width, suggesting that the size of the nanoparticles is very homogenous [89,92].

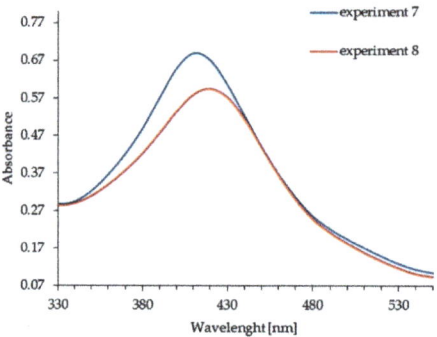

Figure 6. UV-Vis spectra of silver nanoparticles stabilized with 16-6-16 obtained in procedure with excess of reductant.

All UV–Visible absorption spectra exhibit the absorbance peaks in the range of 410–435 nm (Table 3), which is typical feature for colloidal silver nanoparticles [93], confirming our statement that gemini surfactants in the Ag nanodispersion ensures a stabilizing effect on the nanoparticles. These results are in good correlation with what was described previously [87,94,95].

Table 3. The wavelength and absorbance data obtained from UV-Vis spectra.

Experiment	Wavelength [nm]	Absorbance
1	420	0.286
2	435	0.332
3	425	0.359
4	420	0.185
5	420	0.313
6	415	0.397
7	410	0.690
8	419	0.596

Consequently, the colloidal stability is also confirmed by the polydispersity index (PDI), which is given in DLS analysis and should be less or about 0.3. If the PDI is equal or higher than 1, the observable precipitation may occur [96]. The PDI values were between 0.2 and 0.4 in experiments 1–3 and 7–8, illustrating good colloidal features, and PDI values around 0.5 were found to be for the experiments 4–6. Higher PDI values in experiments 4–6 supports relatively broad and slightly distorted width of the UV-Vis spectra (Figure 5b) as well as the visual minor nanodispersions, which were of white color.

3.4. Effect of Spacer Structure on the Size of Silver Nanoparticles

The size of AgNPs capped with gemini and their particle size distributions were obtained using DLS measurements. The Figure 7 shows hydrodynamic diameter of nanoparticles stabilized with gemini surfactants16-6-16 and 16-O-16, which is plotted as a function of the nAg/nGemini molar ratio. The nanoparticle diameter values illustrate an average value of the particle size distribution.

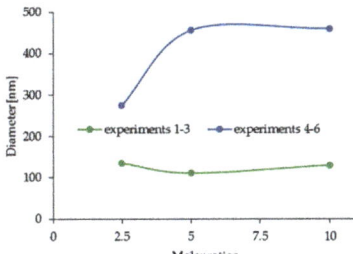

Figure 7. Size of silver nanoparticles capped with 16-O-16 and 16-6-16 as a function of nAg/nGemini molar ratio.

For experiments 4–6 we can observe, that with decreasing of 16-O-16 concentration the silver nanoparticles diameter size increase. However, in the case of 16-6-16 (experiments 1–3) the opposite trend is noticed—we monitor the almost constant diameter size in the range of all molar ratios. Construction of the spacer of gemini surfactants have a considerable impact on nanoparticles size. Compounds with more hydrophobic linker stabilizes more efficiently and gives smaller AgNPs than analog with spacer functionalized by ether group.

Hydrodynamic diameter of silver nanoparticles, stabilized with 16-6-16 from synthetic procedure with excess of reductor, are 110 and 220 nm for experiments 7 and 8, respectively. These results clearly show that synthetic procedure applied in experiment 7 gives the smallest AgNPs with the best stabilizing effect of gemini bilayer. Figure 8 represents the particle size distribution histograms of AgNPs synthesized in the experiments 7 and 8.

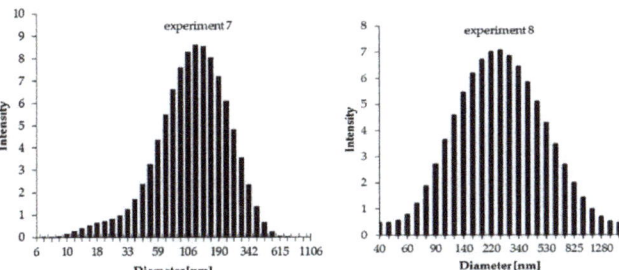

Figure 8. Particle size distribution histograms of AgNPs obtained by DLS measurements.

In fact, the particle sizes of AgNPs measured by DLS is the whole conjugate size, with countslayer of capping agents too [88]. This explains the relatively big diameters. He et al. explained that the gemini cationic surfactants can be adsorbed on the particle surface due to electrostatic interactions between surfactant positive charge head group and counterions attached onto the nanoparticle surface. This causes the formation of coating layers, which lowers the aggregation of particles [90]. Pisárčik et al. demonstrates that gemini surfactants with dodecyl substituent and different spacer length from two to twelve carbon atoms are able to efficiently stabilize silver nanoparticles [87]. Our results indicate that gemini surfactants with longer substituent can also be applied as efficient stabilizer for AgNPs.

Introduction of oxygen atom to spacer results in an increase of hydrophilicity of surfactant molecule. This caused increasing of formed silver nanoparticles size, indicating that the stabilizing effect of gemini bilayer is less efficient. We also demonstrate that, not only the kind of stabilizing agent is crucial, but also synthetic procedures and silver to gemini surfactant molar ratios are very important factors in AgNPs synthesis. These results are in good correlation with what was described previously [87].

4. Conclusions

The 1,6-hexamethylene-bis(N-hexadecyl-N,N-dimethylammonium bromide) and 3-oxa-1,5-pentane-bis(N-hexadecyl-N,N-dimethylammonium) bromide) were obtained via solvent-free green synthetic procedure. The products structures have been confirmed by ^{1}H and ^{13}C NMR, FITR, and elemental analysis. The critical micelle concentrations were specified by conductivity measurement. Silver nanoparticles were obtained via chemical reduction, using gemini surfactants as stabilizing agent. Physicochemical investigations were done by using UV-Vis spectroscopy and DLS measurements. We demonstrated that the structure of the gemini surfactant and conditions of the silver nanoparticles preparation determine their shape and size distribution. It was shown that the hydrophilicity of the dimeric alkylammonium salts can affect the size distribution of AgNPs. The best stabilization, which is represented by a small nanoparticle size, was received in experiment with 16-6-16 as a stabilizing agent, molar ratio nAg/nGemini =5, and with excess of reductant. Results of our research show that application of gemini surfactants is a prospective field of study, which may provide an easy-to-follow green method for the synthesis of nanoparticles.

Author Contributions: Conceptualization, B.B. and A.S.; methodology, A.S. and M.B.; software, M.B.; validation, B.B., A.S., and M.B.; formal analysis, M.B.; investigation, A.S.; resources, B.B.; data curation, A.S.; writing—original draft preparation, A.S.; writing—review and editing, B.B. and M.B.; visualization, M.B.; supervision, B.B.; project administration, B.B.; funding acquisition, B.B. All authors have read and agreed to the published version of the manuscript.

Funding: This paper was funded from the Faculty of Chemistry, Adam Mickiewicz University, subvention for maintaining research potential (B.B.).

Institutional Review Board Statement: Not applicable.

Informed Consent Statement: Not applicable.

Data Availability Statement: The data presented in this study are available in the article.

Conflicts of Interest: The authors declare no conflict of interest.

References

1. Sharma, V.K.; Yngard, R.A.; Lin, Y. Silver nanoparticles: Green synthesis and their antimicrobial activities. *Adv. Colloid Interface Sci.* **2009**, *145*, 83–96. [CrossRef] [PubMed]
2. Helmlinger, J.; Sengstock, C.; Groß-Heitfeld, C.; Mayer, C.; Schildhauer, T.A.; Köller, M.; Epple, M. Silver nanoparticles with different size and shape: Equal cytotoxicity, but different antibacterial effects. *RSC Adv.* **2016**, *6*, 18490–18501. [CrossRef]
3. Ulubayram, K.; Calamak, S.; Shahbazi, R.; Eroglu, I. Nanofibers Based Antibacterial Drug Design, Delivery and Applications. *Curr. Pharm. Des.* **2015**, *21*, 1930–1943. [CrossRef] [PubMed]
4. Ahmed, S.; Ikram, S. Silver Nanoparticles: One Pot Green Synthesis Using Terminalia arjuna Extract for Biological Application. *J. Nanomed. Nanotechnol.* **2015**, *6*. [CrossRef]
5. Yaqoob, S.B.; Adnan, R.; Rameez Khan, R.M.; Rashid, M. Gold, Silver, and Palladium Nanoparticles: A Chemical Tool for Biomedical Applications. *Front. Chem.* **2020**, *8*, 376. [CrossRef]
6. Panáček, A.; Kvítek, L.; Prucek, R.; Kolář, M.; Večeřová, R.; Pizúrová, N.; Sharma, V.K.; Nevěčná, T.; Zbořil, R. Silver Colloid Nanoparticles: Synthesis, Characterization, and Their Antibacterial Activity. *J. Phys. Chem. B* **2006**, *110*, 16248–16253. [CrossRef]
7. Quang, D.V.; Sarawade, P.B.; Hilonga, A.; Kim, J.-K.; Chai, Y.G.; Kim, S.H.; Ryu, J.-Y.; Kim, H.T. Preparation of silver nanoparticle containing silica micro beads and investigation of their antibacterial activity. *Appl. Surf. Sci.* **2011**, *257*, 6963–6970. [CrossRef]
8. Martínez-Gutierrez, F.; Thi, E.P.; Silverman, J.M.; de Oliveira, C.C.; Svensson, S.L.; Hoek, A.V.; Sánchez, E.M.; Reiner, N.E.; Gaynor, E.C.; Pryzdial, E.L.G.; et al. Antibacterial activity, inflammatory response, coagulation and cytotoxicity effects of silver nanoparticles. *Nanomedicine* **2012**, *8*, 328–336. [CrossRef]

9. Salleh, A.; Naomi, R.; Utami, N.D.; Mohammad, A.W.; Mahmoudi, E.; Mustafa, N.; Fauzi, M.B. The Potential of Silver Nanoparticles for Antiviral and Antibacterial Applications: A Mechanism of Action. *Nanomaterials* **2020**, *10*, 1566. [CrossRef] [PubMed]
10. Oluwaniyi, O.O.; Adegoke, H.I.; Adesuji, E.T.; Alabi, A.B.; Bodede, S.O.; Labulo, A.H.; Oseghale, C.O. Biosynthesis of silver nanoparticles using aqueous leaf extract of Thevetia peruviana Juss and its antimicrobial activities. *Appl. Nanosci.* **2016**, *6*, 903–912. [CrossRef]
11. Youssef, F.S.; El-Banna, H.A.; Elzorba, H.Y.; Galal, A.M. Application of some nanoparticles in the field of veterinary medicine. *Int. J. Vet. Sci. Med.* **2019**, *7*, 78–93. [CrossRef] [PubMed]
12. Burduşel, A.-C.; Gherasim, O.; Grumezescu, A.M.; Mogoantă, L.; Ficai, A.; Andronescu, E. Biomedical Applications of Silver Nanoparticles: An Up-to-Date Overview. *Nanomaterials* **2018**, *8*, 681. [CrossRef] [PubMed]
13. Vankar, P.S.; Shukla, D. Biosynthesis of silver nanoparticles using lemon leaves extract and its application for antimicrobial finish on fabric. *Appl. Nanosci.* **2012**, *2*, 163–168. [CrossRef]
14. Xia, Z.-K.; Ma, Q.-H.; Li, S.-Y.; Zhang, D.-Q.; Cong, L.; Tian, Y.-L.; Yang, R.-Y. The antifungal effect of silver nanoparticles on Trichosporon asahii. *J. Microbiol. Immunol. Infect.* **2016**, *49*, 182–188. [CrossRef]
15. Panáček, A.; Kolář, M.; Večeřová, R.; Prucek, R.; Soukupová, J.; Kryštof, V.; Hamal, P.; Zbořil, R.; Kvítek, L. Antifungal activity of silver nanoparticles against Candida spp. *Biomaterials* **2009**, *30*, 6333–6340. [CrossRef]
16. Elgorban, A.M.; El-Samawaty, A.E.-R.M.; Yassin, M.A.; Sayed, S.R.; Adil, S.F.; Elhindi, K.M.; Bakri, M.; Khan, M. Antifungal silver nanoparticles: Synthesis, characterization and biological evaluation. *Biotechnol. Biotechnol. Equip.* **2016**, *30*, 56–62. [CrossRef]
17. Usman, M.; Farooq, M.; Wakeel, A.; Nawaz, A.; Cheema, S.A.; Rehman, H.; Ashraf, I.; Sanaullah, M. Nanotechnology in agriculture: Current status, challenges and future opportunities. *Sci. Total Environ.* **2020**, *721*, 137778. [CrossRef]
18. Galdiero, S.; Falanga, A.; Vitiello, M.; Cantisani, M.; Marra, V.; Galdiero, M. Silver Nanoparticles as Potential Antiviral Agents. *Molecules* **2011**, *16*, 8894–8918. [CrossRef]
19. Thi Ngoc Dung, T.; Nang Nam, V.; Thi Nhan, T.; Ngoc, T.T.B.; Minh, L.Q.; Nga, B.T.T.; Phan Le, V.; Viet Quang, D. Silver nanoparticles as potential antiviral agents against African swine fever virus. *Mater. Res. Express* **2020**, *6*, 1250–1259. [CrossRef]
20. Mori, Y.; Ono, T.; Miyahira, Y.; Nguyen, V.Q.; Matsui, T.; Ishihara, M. Antiviral activity of silver nanoparticle/chitosan composites against H1N1 influenza A virus. *Nanoscale Res. Lett.* **2013**, *8*, 93. [CrossRef]
21. Bekele, A.Z.; Gokulan, K.; Williams, K.M.; Khare, S. Dose and Size-Dependent Antiviral Effects of Silver Nanoparticles on Feline Calicivirus, a Human Norovirus Surrogate. *Foodborne Pathog. Dis.* **2016**, *13*, 239–244. [CrossRef] [PubMed]
22. Singh, L.; Kruger, H.G.; Maguire, G.E.M.; Govender, T.; Parboosing, R. The role of nanotechnology in the treatment of viral infections. *Ther. Adv. Infect. Dis.* **2017**, *4*, 105–131. [CrossRef] [PubMed]
23. Seino, S.; Imoto, Y.; Kosaka, T.; Nishida, T.; Nakagawa, T.; Yamamoto, T.A. Antiviral Activity of Silver Nanoparticles Immobilized onto Textile Fabrics Synthesized by Radiochemical Process. *MRS Adv.* **2016**, *1*, 705–710. [CrossRef]
24. Balagna, C.; Perero, S.; Percivalle, E.; Nepita, E.V.; Ferraris, M. Virucidal effect against coronavirus SARS-CoV-2 of a silver nanocluster/silica composite sputtered coating. *Open Ceram.* **2020**, *1*, 100006. [CrossRef]
25. Nikaeen, G.; Abbaszadeh, S.; Yousefinejad, S. Application of nanomaterials in treatment, anti-infection and detection of coronaviruses. *Nanomedicine* **2020**, *15*, 1501–1512. [CrossRef] [PubMed]
26. Lee, S.; Jun, B.-H. Silver Nanoparticles: Synthesis and Application for Nanomedicine. *Int. J. Mol. Sci.* **2019**, *20*, 865. [CrossRef]
27. Sotiriou, G.A.; Pratsinis, S.E. Engineering nanosilver as an antibacterial, biosensor and bioimaging material. *Curr. Opin. Chem. Eng.* **2011**, *1*, 3–10. [CrossRef]
28. Marassi, V.; Cristo, L.D.; Smith, G.J.; Ortelli, S.; Blosi, M.; Costa, A.L.; Reschiglian, P.; Volkov, Y.; Prina-Mello, A. Silver nanoparticles as a medical device in healthcare settings: A five-step approach for candidate screening of coating agents. *R. Soc. Open Sci.* **2018**, *5*, 171113. [CrossRef]
29. Liu, J.; Zhao, Y.; Guo, Q.; Wang, Z.; Wang, H.; Yang, Y.; Huang, Y. TAT-modified nanosilver for combating multidrug-resistant cancer. *Biomaterials* **2012**, *33*, 6155–6161. [CrossRef]
30. Ardhendu, K.M. Silver Nanoparticles as Drug Delivery Vehicle. *Glob. J. Nanomed.* **2017**, *3*, 555607. [CrossRef]
31. Lee, K.-S.; El-Sayed, M.A. Gold and Silver Nanoparticles in Sensing and Imaging: Sensitivity of Plasmon Response to Size, Shape, and Metal Composition. *J. Phys. Chem. B* **2006**, *110*, 19220–19225. [CrossRef] [PubMed]
32. Wang, L.; Sun, Y.; Che, G.; Li, Z. Self-assembled silver nanoparticle films at an air–liquid interface and their applications in SERS and electrochemistry. *App. Surf. Sci.* **2011**, *257*, 7150–7155. [CrossRef]
33. Alvarez-Puebla, R.A.; Aroca, R.F. Synthesis of Silver Nanoparticles with Controllable Surface Charge and Their Application to Surface-Enhanced Raman Scattering. *Anal. Chem.* **2009**, *81*, 2280–2285. [CrossRef] [PubMed]
34. Yang, F.; Wang, Q.; Gu, Z.; Fang, K.; Marriott, G.; Gu, N. Silver Nanoparticle-Embedded Microbubble as a Dual-Mode Ultrasound and Optical Imaging Probe. *ACS Appl. Mater. Interfaces* **2013**, *5*, 9217–9223. [CrossRef] [PubMed]
35. Varghese Alex, K.; Tamil Pavai, P.; Rugmini, R.; Shiva Prasad, M.; Kamakshi, K.; Sekhar, K.C. Green Synthesized Ag Nanoparticles for Bio-Sensing and Photocatalytic Applications. *ACS Omega* **2020**, *5*, 13123–13129. [CrossRef]
36. Manno, D.; Filippo, E.; Di Giulio, M.; Serra, A. Synthesis and characterization of starch-stabilized Ag nanostructures for sensors applications. *J. Non-Cryst. Solids* **2008**, *354*, 5515–5520. [CrossRef]
37. Dong, X.-Y.; Gao, Z.-W.; Yang, K.-F.; Zhang, W.-Q.; Xu, L.-W. Nanosilver as a new generation of silver catalysts in organic transformations for efficient synthesis of fine chemicals. *Catal. Sci. Technol.* **2015**, *5*, 2554–2574. [CrossRef]

38. Treshchalov, A.; Erikson, H.; Puust, L.; Tsarenko, S.; Saar, R.; Vanetsev, A.; Tammeveski, K.; Sildos, I. Stabilizer-free silver nanoparticles as efficient catalysts for electrochemical reduction of oxygen. *J. Colloid Interface Sci.* **2017**, *491*, 358–366. [CrossRef]
39. Yaqoob, A.A.; Umar, K.; Ibrahim, M.N.M. Silver nanoparticles: Various methods of synthesis, size affecting factors and their potential applications—A review. *Appl. Nanosci.* **2020**, *10*, 1369–1378. [CrossRef]
40. Zhang, X.-F.; Liu, Z.-G.; Shen, W.; Gurunathan, S. Silver Nanoparticles: Synthesis, Characterization, Properties, Applications, and Therapeutic Approaches. *Int. J. Mol. Sci.* **2016**, *17*, 1534. [CrossRef]
41. Abou El-Nour, K.M.M.; Eftaiha, A.; Al-Warthan, A.; Ammar, R.A.A. Synthesis and applications of silver nanoparticles. *Arab. J. Chem.* **2010**, *3*, 135–140. [CrossRef]
42. Chouhan, N. Silver Nanoparticles: Synthesis, Characterization and Applications. In *Silver Nanoparticles—Fabrication, Characterization and Applications*; Maaz, K., Ed.; InTech: Rijeka, Croatia, 2018; ISBN 978-1-78923-478-7.
43. Khodashenas, B.; Ghorbani, H.R. Synthesis of silver nanoparticles with different shapes. *Arab. J. Chem.* **2019**, *12*, 1823–1838. [CrossRef]
44. AL-Thabaiti, S.A.; Al-Nowaiser, F.M.; Obaid, A.Y.; Al-Youbi, A.O.; Khan, Z. Formation and characterization of surfactant stabilized silver nanoparticles: A kinetic study. *Colloids Surf. B* **2008**, *67*, 230–237. [CrossRef] [PubMed]
45. Menger, F.M.; Littau, C.A. Gemini-surfactants: Synthesis and properties. *J. Am. Chem. Soc.* **1991**, *113*, 1451–1452. [CrossRef]
46. Alami, E.; Beinert, G.; Marie, P.; Zana, R. Alkanediyl-α,ω-bis(dimethylalkylammonium bromide) surfactants. 3. Behavior at the air-water interface. *Langmuir* **1993**, *9*, 1465–1467. [CrossRef]
47. Brycki, B.; Kowalczyk, I.; Kozirog, A. Synthesis, Molecular Structure, Spectral Properties and Antifungal Activity of Polymethylene-α,ω-bis(N,N- dimethyl-N-dodecyloammonium Bromides). *Molecules* **2011**, *16*, 319–335. [CrossRef]
48. Brycki, B.E.; Kowalczyk, I.H.; Szulc, A.; Kaczerewska, O.; Pakiet, M. Multifunctional Gemini Surfactants: Structure, Synthesis, Properties and Applications. In *Application and Characterization of Surfactants*; Najjar, R., Ed.; InTech: Rijeka, Croatia, 2017; ISBN 978-953-51-3325-4.
49. Menger, F.M.; Keiper, J.S. Gemini Surfactants. *Angew. Chem. Int. Ed.* **2000**, *39*, 1906–1920. [CrossRef]
50. Brycki, B.; Szulc, A.; Koenig, H.; Kowalczyk, I.; Pospieszny, T.; Górka, S. Effect of the alkyl chain length on micelle formation for bis(N-alkyl-N,N-dimethylethylammonium)ether dibromides. *C. R. Chim.* **2019**, *22*, 386–392. [CrossRef]
51. Chlebicki, J.; Węgrzyńska, J.; Wilk, K.A. Surface-active, micellar, and antielectrostatic properties of bis-ammonium salts. *J. Colloid Interface Sci.* **2008**, *323*, 372–378. [CrossRef]
52. Das, S.; Mukherjee, I.; Paul, B.K.; Ghosh, S. Physicochemical Behaviors of Cationic Gemini Surfactant (14-4-14) Based Microheterogeneous Assemblies. *Langmuir* **2014**, *30*, 12483–12493. [CrossRef]
53. Pisárčik, M.; Devínsky, F. Surface tension study of cationic gemini surfactants binding to DNA. *Cent. Eur. J. Chem.* **2014**, *12*, 577–585. [CrossRef]
54. Devínsky, F.; Masárová, Ľ.; Lacko, I. Surface activity and micelle formation of some new bisquaternary ammonium salts. *J. Colloid Interface Sci.* **1985**, *105*, 235–239. [CrossRef]
55. Kowalczyk, I.; Pakiet, M.; Szulc, A.; Koziróg, A. Antimicrobial Activity of Gemini Surfactants with Azapolymethylene Spacer. *Molecules* **2020**, *25*, 4054. [CrossRef] [PubMed]
56. Koziróg, A.; Brycki, B. Monomeric and gemini surfactants as antimicrobial agents—Influence on environmental and reference strains. *Acta Biochim. Pol.* **2015**, *62*, 879–883. [CrossRef] [PubMed]
57. Kaczerewska, O.; Leiva-Garcia, R.; Akid, R.; Brycki, B.; Kowalczyk, I.; Pospieszny, T. Heteroatoms and π electrons as favorable factors for efficient corrosion protection. *Mater. Corros.* **2019**, *70*, 1099–1110. [CrossRef]
58. Garcia, M.T.; Kaczerewska, O.; Ribosa, I.; Brycki, B.; Materna, P.; Drgas, M. Biodegradability and aquatic toxicity of quaternary ammonium-based gemini surfactants: Effect of the spacer on their ecological properties. *Chemosphere* **2016**, *154*, 155–160. [CrossRef]
59. Kaczerewska, O.; Brycki, B.; Ribosa, I.; Comelles, F.; Garcia, M.T. Cationic gemini surfactants containing an O-substituted spacer and hydroxyethyl moiety in the polar heads: Self-assembly, biodegradability and aquatic toxicity. *J. Ind. Eng. Chem.* **2018**, *59*, 141–148. [CrossRef]
60. Brycki, B.; Waligórska, M.; Szulc, A. The biodegradation of monomeric and dimeric alkylammonium surfactants. *J. Hazard. Mater.* **2014**, *280*, 797–815. [CrossRef]
61. Akram, M.; Anwar, S.; Ansari, F.; Bhat, I.A.; Kabir-ud-Din, K.-D. Bio-physicochemical analysis of ethylene oxide-linked diester-functionalized green cationic gemini surfactants. *RSC Adv.* **2016**, *6*, 21697–21705. [CrossRef]
62. Bergero, M.F.; Liffourrena, A.S.; Opizzo, B.A.; Fochesatto, A.S.; Lucchesi, G.I. Immobilization of a microbial consortium on Ca-alginate enhances degradation of cationic surfactants in flasks and bioreactor. *Int. Biodeterior. Biodegrad.* **2017**, *117*, 39–44. [CrossRef]
63. Mondal, M.H.; Roy, A.; Malik, S.; Ghosh, A.; Saha, B. Review on chemically bonded geminis with cationic heads: Second-generation interfactants. *Res. Chem. Intermed.* **2016**, *42*, 1913–1928. [CrossRef]
64. Mondal, M.H.; Malik, S.; Roy, A.; Saha, R.; Saha, B. Modernization of surfactant chemistry in the age of gemini and bio-surfactants: A review. *RSC Adv.* **2015**, *5*, 92707–92718. [CrossRef]
65. Martín, V.I.; de la Haba, R.R.; Ventosa, A.; Congiu, E.; Ortega-Calvo, J.J.; Moyá, M.L. Colloidal and biological properties of cationic single-chain and dimeric surfactants. *Colloids Surf. B* **2014**, *114*, 247–254. [CrossRef] [PubMed]

66. Brycki, B.; Szulc, A. Gemini Alkyldeoxy-D-Glucitolammonium Salts as Modern Surfactants and Microbiocides: Synthesis, Antimicrobial and Surface Activity, Biodegradation. *PLoS ONE* **2014**, *9*, e84936. [CrossRef]
67. Dani, U.; Bahadur, A.; Kuperkar, K. Micellization, Antimicrobial Activity and Curcumin Solubilization in Gemini Surfactants: Influence of Spacer and Non-Polar Tail. *Colloids Interface Sci. Commun.* **2018**, *25*, 22–30. [CrossRef]
68. Jennings, M.C.; Buttaro, B.A.; Minbiole, K.P.C.; Wuest, W.M. Bioorganic Investigation of Multicationic Antimicrobials to Combat QAC-Resistant *Staphylococcus aureus*. *ACS Infect. Dis.* **2015**, *1*, 304–309. [CrossRef]
69. Zhang, S.; Ding, S.; Yu, J.; Chen, X.; Lei, Q.; Fang, W. Antibacterial Activity, *in Vitro* Cytotoxicity, and Cell Cycle Arrest of Gemini Quaternary Ammonium Surfactants. *Langmuir* **2015**, *31*, 12161–12169. [CrossRef]
70. Kuperkar, K.; Modi, J.; Patel, K. Surface-Active Properties and Antimicrobial Study of Conventional Cationic and Synthesized Symmetrical Gemini Surfactants. *J. Surfactants Deterg.* **2012**, *15*, 107–115. [CrossRef]
71. Obłąk, E.; Piecuch, A.; Krasowska, A.; Łuczyński, J. Antifungal activity of gemini quaternary ammonium salts. *Microbiol. Res.* **2013**, *168*, 630–638. [CrossRef]
72. Koziróg, A.; Kręgiel, D.; Brycki, B. Action of Monomeric/Gemini Surfactants on Free Cells and Biofilm of Asaia lannensis. *Molecules* **2017**, *22*, 2036. [CrossRef]
73. Labena, A.; Hegazy, M.A.; Sami, R.M.; Hozzein, W.N. Multiple Applications of a Novel Cationic Gemini Surfactant: Anti-Microbial, Anti-Biofilm, Biocide, Salinity Corrosion Inhibitor, and Biofilm Dispersion (Part II). *Molecules* **2020**, *25*, 1348. [CrossRef] [PubMed]
74. Kumar, N.; Tyagi, R. Industrial Applications of Dimeric Surfactants: A Review. *J. Dispers. Sci. Technol.* **2014**, *35*, 205–214. [CrossRef]
75. Brycki, B.; Drgas, M.; Bielawska, M.; Zdziennicka, A.; Jańczuk, B. Synthesis, spectroscopic studies, aggregation and surface behavior of hexamethylene-1,6-bis(N,N-dimethyl-N-dodecylammonium bromide). *J. Mol. Liq.* **2016**, *221*, 1086–1096. [CrossRef]
76. Han, Y.; Wang, Y. Aggregation behavior of gemini surfactants and their interaction with macromolecules in aqueous solution. *Phys. Chem. Chem. Phys.* **2011**, *13*, 1939–1956. [CrossRef]
77. Zana, R. Dimeric (Gemini) Surfactants: Effect of the Spacer Group on the Association Behavior in Aqueous Solution. *J. Colloid Interface Sci.* **2002**, *248*, 203–220. [CrossRef]
78. Singh, V.; Tyagi, R. Unique Micellization and CMC Aspects of Gemini Surfactant: An Overview. *J. Dispers. Sci. Technol.* **2014**, *35*, 1774–1792. [CrossRef]
79. ud din Parray, M.; Maurya, N.; Ahmad Wani, F.; Borse, M.S.; Arfin, N.; Ahmad Malik, M.; Patel, R. Comparative effect of cationic gemini surfactant and its monomeric counterpart on the conformational stability of phospholipase A2. *J. Mol. Struct.* **2019**, *1175*, 49–55. [CrossRef]
80. Egorova, E.M.; Kaba, S.I. The effect of surfactant micellization on the cytotoxicity of silver nanoparticles stabilized with aerosol-OT. *Toxicol. In Vitro* **2019**, *57*, 244–254. [CrossRef]
81. Zana, R. Critical Micellization Concentration of Surfactants in Aqueous Solution and Free Energy of Micellization. *Langmuir* **1996**, *12*, 1208–1211. [CrossRef]
82. Menger, F.M.; Littau, C.A. Gemini surfactants: A new class of self-assembling molecules. *J. Am. Chem. Soc.* **1993**, *115*, 10083–10090. [CrossRef]
83. Zana, R.; Levy, H. Alkanediyl-α,ω-bis(dimethylalkylammonium bromide) surfactants (dimeric surfactants) Part 6. CMC of the ethanediyl- 1,2-bis(dimethylalkylammonium bromide) series. *Colloids Surf. A* **1997**, *127*, 229–232. [CrossRef]
84. Dam, T.; Engberts, J.B.F.N.; Karthäuser, J.; Karaborni, S.; van Os, N.M. Synthesis, surface properties and oil solubilisation capacity of cationic gemini surfactants. *Colloids Surf. A* **1996**, *118*, 41–49. [CrossRef]
85. Oda, R.; Candau, S.J.; Oda, R.; Huc, I. Gemini surfactants, the effect of hydrophobic chain length and dissymmetry. *Chem. Commun.* **1997**, 2105–2106. [CrossRef]
86. Chang, H.; Cui, Y.; Wang, Y.; Li, G.; Gao, W.; Li, X.; Zhao, X.; Wei, W. Wettability and adsorption of PTFE and paraffin surfaces by aqueous solutions of biquaternary ammonium salt Gemini surfactants with hydroxyl. *Colloids Surf. A* **2016**, *506*, 416–424. [CrossRef]
87. Pisárčik, M.; Jampílek, J.; Lukáč, M.; Horáková, R.; Devínsky, F.; Bukovský, M.; Kalina, M.; Tkacz, J.; Opravil, T. Silver Nanoparticles Stabilised by Cationic Gemini Surfactants with Variable Spacer Length. *Molecules* **2017**, *22*, 1794. [CrossRef]
88. Siddiq, A.M.; Parandhaman, T.; Begam, A.F.; Das, S.K.; Alam, M.S. Effect of gemini surfactant (16-6-16) on the synthesis of silver nanoparticles: A facile approach for antibacterial application. *Enzym. Microb. Technol.* **2016**, *95*, 118–127. [CrossRef]
89. Xu, J.; Han, X.; Liu, H.; Hu, Y. Synthesis and optical properties of silver nanoparticles stabilized by gemini surfactant. *Colloids Surf. A* **2006**, *273*, 179–183. [CrossRef]
90. He, S.; Chen, H.; Guo, Z.; Wang, B.; Tang, C.; Feng, Y. High-concentration silver colloid stabilized by a cationic gemini surfactant. *Colloids Surf. A* **2013**, *429*, 98–105. [CrossRef]
91. Guzman, M.; Dille, J.; Godet, S. Synthesis and antibacterial activity of silver nanoparticles against gram-positive and gram-negative bacteria. *Nanomedicine* **2012**, *8*, 37–45. [CrossRef]
92. He, S.; Yao, J.; Jiang, P.; Shi, D.; Zhang, H.; Xie, S.; Pang, S.; Gao, H. Formation of Silver Nanoparticles and Self-Assembled Two-Dimensional Ordered Superlattice. *Langmuir* **2001**, *17*, 1571–1575. [CrossRef]
93. Njagi, E.C.; Huang, H.; Stafford, L.; Genuino, H.; Galindo, H.M.; Collins, J.B.; Hoag, G.E.; Suib, S.L. Biosynthesis of Iron and Silver Nanoparticles at Room Temperature Using Aqueous Sorghum Bran Extracts. *Langmuir* **2011**, *27*, 264–271. [CrossRef] [PubMed]

94. Li, D.; Fang, W.; Feng, Y.; Geng, Q.; Song, M. Stability properties of water-based gold and silver nanofluids stabilized by cationic gemini surfactants. *J. Taiwan Inst. Chem. Eng.* **2019**, *97*, 458–465. [CrossRef]
95. Siddiq, A.M.; Thangam, R.; Madhan, B.; Alam, M.S. Counterion coupled (COCO) gemini surfactant capped Ag/Au alloy and core–shell nanoparticles for cancer therapy. *RSC Adv.* **2019**, *9*, 37830–37845. [CrossRef]
96. Haque, M.N.; Kwon, S.; Cho, D. Formation and stability study of silver nano-particles in aqueous and organic medium. *Korean J. Chem. Eng.* **2017**, *34*, 2072–2078. [CrossRef]

Article

Gemini Surfactant as a Template Agent for the Synthesis of More Eco-Friendly Silica Nanocapsules

Olga Kaczerewska [1,*], Isabel Sousa [1], Roberto Martins [2], Joana Figueiredo [2], Susana Loureiro [2] and João Tedim [1]

1. CICECO-Aveiro Institute of Materials and Department of Materials and Ceramic Engineering, University of Aveiro, 3810-193 Aveiro, Portugal; isabel.sa.correia@ua.pt (I.S.); joao.tedim@ua.pt (J.T.)
2. CESAM-Centre for Environmental and Marine Studies and Department of Biology, University of Aveiro, 3810-193 Aveiro, Portugal; roberto@ua.pt (R.M.); jrmf@ua.pt (J.F.); sloureiro@ua.pt (S.L.)
* Correspondence: olga.kaczerewska@ua.pt

Received: 14 October 2020; Accepted: 13 November 2020; Published: 15 November 2020

Abstract: Silica mesoporous nanocapsules are a class of "smart" engineered nanomaterials (ENMs) applied in several fields. Recent studies have highlighted that they can exert deleterious effects into marine organisms, attributed to the use of the toxic cationic surfactant N-hexadecyl -N,N,N-trimethylammonium bromide (CTAB) during the synthesis of ENMs. The present study reports the successful synthesis and characterization of novel gemini surfactant-based silica nanocapsules. The gemini surfactant 1,4-bis-[N-(1-dodecyl)-N,N-dimethylammoniummethyl]benzene dibromide (QSB2-12) was chosen as a more environmentally-friendly replacement of CTAB. Nanocapsules were characterized by scanning electron microscopy (SEM), Fourier-transformed infrared spectroscopy (FTIR), dynamic light scattering (DLS), thermogravimetric analysis (TGA) and N_2 adsorption-desorption isotherms. Short-term exposure effects of new ENMs were evaluated in four marine species (*Nannochloropsis gaditana*, *Tetraselmis chuii* and *Phaeodactylum tricornutum*) and the microcrustacean (*Artemia salina*). The replacement of the commercial cationic surfactant by the gemini surfactant does not change the structure nor the environmental behaviour in seawater of the newly synthesised silica nanocontainers. Additionally, it is demonstrated that using gemini surfactants can reduce the toxicity of novel silica nanocapsules towards the tested marine species. As a result, environmentally-friendly ENMs can be obtained based on a safe-by-design approach, thereby fitting the concept of Green Chemistry.

Keywords: cationic surfactants; microemulsion; mesoporous silica; encapsulation; ecotoxicity

1. Introduction

Engineered nanomaterials (ENMs) have been proposed as new solutions for encapsulation of active compounds, such as corrosion inhibitors, antifouling agents, pH indicators, drugs or dyes [1–7]. Nanocapsules are core–shell structures with diameter generally ranging between 100–500 nm [8], and one of the most used nanocapsules are based on mesoporous silica, typically with a diameter of 100–200 nm [9]. This cutting-edge immobilization technique has assumed a relevant role in the field of smart coatings, by providing a controlled release of encapsulated molecules, reducing their toxicity (compared to the free forms), preventing direct interaction with coating matrices, reducing leakages and ultimately increasing the coating's service life [3,10,11]. However, state-of-the-art information shows that even empty silica nanocapsules show some degree of toxicity [10]. They have been found to be more toxic towards selected marine species than other engineered nanomaterials, namely Zn-Al layered double hydroxides (LDH) [12,13]. Recent studies have revealed that the silica nano-form or the formation of agglomerates in artificial saltwater, over time, is not necessarily the source of silica nanocapsules toxicity [10–12,14]. Indeed, it is related to the presence of a cationic surfactant, *N*-hexadecyl-*N*,*N*,*N*-trimethylammonium bromide (CTAB) which is used as a template during nanomaterials' synthesis. Replacing CTAB with a greener surfactant has been suggested previously in order to reduce capsules' toxicity [10]. There are strategies that allow the complete removal of CTAB from capsules' surface such as extraction with solvents solutions, washing with a solution of ethanolic HCl or thermal treatment (calcination) [9,15,16]. Although these steps are effective for removal of synthesis remnants from empty silica nanocapsules, they may be undesirable if the entrapment of active molecules inside the nanomaterials is the main goal: calcination leads to complete degradation of all organic molecules, while using other solvents may lead to leaching of the encapsulated agents as well. Therefore, replacing CTAB with a more eco-friendly surfactant would prevent toxicological issues and may avoid the implementation of additional steps.

Cationic gemini surfactants are a group of innovative quaternary ammonium salts that consist of two monomeric moieties linked by a spacer. Each moiety is made of a hydrophilic part (positively charged nitrogen atom) and a hydrophobic part (alkyl chain) [17]. Due to this dimeric structure, gemini surfactants show unique surface properties, such as critical micelle concentrations (CMC), which is lower comparing with conventional cationic surfactants, with higher efficacy in lowering surface tension (for surfactants with the same number of carbon atoms in hydrophobic parts). CMC is the concentration at which micelles start forming [18], and the ability of cationic surfactants to form stable oil in water emulsion allows their use as a template in silica nanocapsules synthesis [14]. For a conventional monomeric surfactant, *N*-dodecyl-*N*,*N*,*N*-trimethylammonium bromide (DTAB), the CMC is 16 mM whereas for a dimeric surfactant, with a rigid spacer, such as 1,4-bis-[*N*-(1-dodecyl)-*N*,*N*-dimethylammoniummethyl]benzene dibromide (QSB2-12) the CMC is 1.21 mM [18]. For CTAB the CMC is 0.907 mM [19], lower than for QSB2-12. It has been reported that by modifying the surfactants' structure, aggregation behavior may be changed. Increasing the number of hydrophilic parts decreases the CMC (when the hydrophobic part is fixed), which is also observed when an increase in the number of carbon chains of the hydrophobic part occurs (when hydrophilic part is fixed) [17].

Gemini surfactants are known to be less toxic to freshwater organisms [20]. As an example, and looking at the key freshwater species *Daphnia magna*, CTAB was found to be 28 times more toxic than QSB2-12: LC_{50} of 0.026 mg/L [21] vs. 0.73 mg/L [20], respectively. Comparing QSB2-12 with its monomeric analogue DTAB, the gemini surfactant is also less toxic than the conventional one (LC_{50} for DTAB towards *Daphnia magna* is 0.35 mg/L [20]). It has been already reported that elongating an alkyl chain leads to an increase in toxicity [22,23]. As an example, for betainate cationic dimeric surfactants, LC_{50} values towards *Daphnia magna* decreases from 50 to 7.5 mg/L [22] when the number of carbon atoms in the hydrophobic parts increases from 8 to 10. It can be also observed for the above-mentioned monomeric surfactants DTAB (C12) and CTAB (C16). The compound with 12 carbon atoms is classified as very toxic toward *Daphnia magna* [20,24] whereas the one with 16 carbon atoms as extremely toxic [21]. Nevertheless, when we look at marine or estuarine organisms, not many information is available regarding gemini surfactant toxicity. This represents a serious gap in marine hazard information available, as nanocapsules synthetized using gemini surfactants are often used in maritime infrastructures. One of the few examples is a study where gemini surfactants were shown to be 20-times less toxic when looking at marine microalgae than CTAB [25].

Taking into consideration the relation between surfactant structure, aggregation properties and ecotoxicity, environmentally-friendly silica nanocapsules may be designed. This approach, named safe-by-design (SbD), focuses on setting, at minimum, any sources of possible hazards for the environment and humans. SbD aims at ecologically oriented design since the beginning, meaning that all possible safety issues will be identified and replaced with non-toxic or less toxic alternatives [26,27].

Therefore, the aim of this study was to develop a safe-by-design approach to silica nanocapsules synthesis. In order to achieve it, CTAB was replaced by the gemini surfactant QSB2-12 as a template for the nanocapsules' synthesis. The nanocontainers based on both cationic surfactants were fully characterized by scanning electron microscopy (SEM), Fourier-transformed infrared spectroscopy (FTIR), dynamic light scattering (DLS), thermogravimetric analysis (TG/DTA) and N_2 adsorption–desorption isotherms, as well as, in terms of their ecotoxicity towards chosen marine microalgae and crustaceans. The marine environment is the compartment where most of these ENMs will be applied and, therefore, looking at effects to marine biota is crucial and it will help to fill in the gap on the few available information regarding hazard assessment.

2. Materials and Methods

2.1. Materials

Ammonia solution (NH_4OH) (25%) was purchased from Merck (Portugal). Hexadecyltrimethylammonium bromide (CTAB) (≥98%), tetraethyl orthosilicate (TEOS) (≥99%), diethyl ether and sodium chloride (NaCl) were purchased from Sigma-Aldrich (Oeiras, Portugal).

Cationic gemini surfactant 1,4-bis-[N-(1-dodecyl)-N,N-dimethylammoniummethyl]benzene dibromide (QSB2-12) was synthesised according to the methods described in the literature [20].

The structures of cationic surfactants are presented in Figure 1.

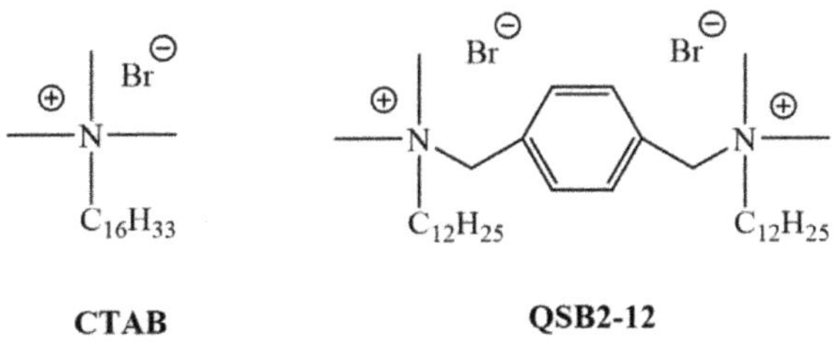

Figure 1. Structure of the cationic surfactants.

2.2. Methods

2.2.1. Nanocapsules Synthesis and Characterization

Silica nanocapsules (SiNC) were synthesized in a one-step process through an oil-in-water microemulsion (Figure 2) based on a published procedure, with minor alterations [9]. Diethyl ether was used as a co-solvent, ammonia as a catalyst and TEOS as a silica precursor. The water phase consisted of different amounts of cationic surfactant (Figure 2) dissolved in 35 mL water and 0.25 mL of ammonia solution. Diethyl ether (25 mL) was then added (dropwise) to the water phase and an oil-in-water microemulsion was obtained. Subsequently, 2 mL of TEOS was added dropwise to the microemulsion under controlled stirring and kept in a closed vessel for 24 h. The obtained products were filtered, washed with warm water (~40 °C) and dried at 60 °C.

Capsules morphology was characterized by scanning electron microscopy (SEM) coupled with energy dispersive spectroscopy (EDS) (Hitachi SU-70 electron microscope) and by scanning transmission electron microscopy (STEM) with secondary electron (SE) imagining capability (Hitachi STEM/SEM HD-2700).

Fourier-transformed infrared spectroscopy (FTIR) was used to identify characteristic peaks of the synthesized nanomaterials. FTIR analysis was performed on a Bruker tensor 27 spectrophotometer coupled with an ATR device.

A Malvern Zetasizer Nano-ZS instrument was used to perform dynamic light scattering (DLS) and zeta potential (ζ) measurements. The measurements were performed in deionized water and the concentration of silica nanocapsules was 10 mg/L.

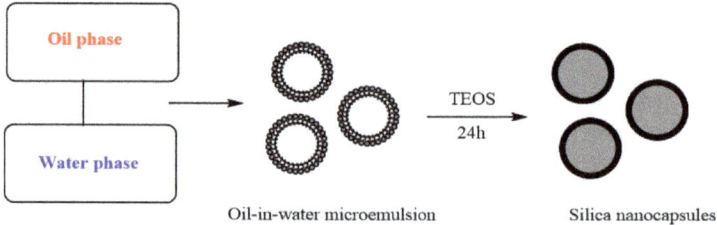

Surfactant	CTAB	QSB2-12
Mass used	0.25 g (0.67 mmol)	0.20 g (0.29 mmol)

Figure 2. Scheme for the silica nanocapsules synthesis and a quantity of surfactants used.

Thermogravimetric analysis (TG/DTA) was conducted on a *Sataram-Labsys* system under air atmosphere, with a heating rate of 10 °C min^{-1} from room temperature up to 800 °C. A portion of these nanocapsules was calcined at 550 °C for 5 h, with a heating rate of 10 °C min^{-1}.

Textural properties of nanocapsules were evaluated based on the adsorption–desorption isotherms of N_2 at 196 °C, performed on the equipment Quantachrome NOVA 4200e. Samples were previously degassed at 180 °C for 6 h. The specific area (S_{BET}) was calculated by the BET method (Braunauer, Emmett and Teller), the total pore volume ($V_{Pp/p0=0.98}$) was obtained from the volume of N_2 adsorbed at $p/p^0 = 0.98$. The most frequent pore diameters ($\varnothing_{pores}$) was calculated by the BJH method (Barrett–Joyner–Halenda), applied to the desorption branch of the isotherm [14].

2.2.2. Ecotoxicity Tests

The short-term toxic effects of the two silica nanocapsules were evaluated on marine microalgae species (*Nannochloropsis gaditana, Phaeodactylum tricornutum, Tetraselmis chuii*) and on a microcrustacean species (*Artemia salina*), following the standard protocols OECD 201 (2011) [28] and ISO 10253 (2016) [29], respectively, with some adaptations fully described by Kaczerewska et al. [25] Briefly, tests were run with 0.45 μm filtered artificial seawater (ASW) and for each compound, five concentrations plus one negative control (ASW only) were tested (n = 4 for microalgae; n = 3 for crustaceans). Range-finding tests were run for exposure concentrations ranging from 0.01 mg/L to 100 mg/L. Definitive exposure tests included the following exposure concentrations: *Nanochloropsis gaditana* and *Tetraselmis chuii*: 2 mg/L, 4 mg/L, 6 mg/L, 8 mg/L, 10 mg/L; *Phaeodactylum tricornutum*: 2 mg/L, 4 mg/L, 6 mg/L, 8 mg/L, 10 mg/L for SiNC_QSB2-12 and 0.5 mg/L, 1 mg/L, 2 mg/L, 4 mg/L, 6 mg/L for SiNC_CTAB; *Artemia salina*: 6.25 mg/L, 12.5 mg/L, 25 mg/L, 50 mg/L, 100 mg/L. Microalgae growth inhibition was monitored for 72 h through fluorescence daily measurements whereas crustaceans mortality or immobilization was checked after 48 h of exposure. The median lethal (LC_{50}) and median growth inhibition (IC_{50}) concentrations were determined by a non-linear regression model with the software *Graphpad Prism* v.6.0. Then, toxicity endpoints were categorized according to the EC Directive 93/67/EEC scheme adapted by Blaise et al. [30] for nanomaterials: non-toxic (L/IC_{50} > 100 mg/L), harmful (10 > L/IC_{50} ≥ 100 mg/L), toxic (1 > L/IC_{50} ≥ 10 mg/L), very toxic (0.1 > L/IC_{50} ≥ 1 mg/L) and extremely toxic (L/IC_{50} ≤ 0.1 mg/L).

l/LC_{50} values estimated for both nanomaterials were statistically compared. A table containing the logL/IC_{50}, associated standard error and degrees of freedom (data extracted from each non-linear regression report) for both SiNC (QSB2-12) and SiNC (CTAB) was prepared for each tested species. Then, the null hypothesis that both best fitting datasets were similar was tested ($p < 0.05$) through a *t*-test, followed by an F-test to compare the data variance (total of 4 independent tests).

3. Results and Discussion

3.1. Synthesis

As mentioned in the Materials section, silica nanocapsules have been obtained in a one-step process with minor adjustments. Both SiNC_CTAB and SiNC_QSB2-12 were obtained as white powders. CTAB concentration used for the synthesis was 19 mM, which is definitely above its CMC (0.907 mM). The number of moles used for QSB2-12 was reduced by half (11.6 mM) due to its dimeric structure. Zeta potential of CTAB and QSB2-12 microemulsions were +58.3 mV and +41.9 mV, respectively. Values of ζ for the surfactants differ slightly suggesting similar stability of the obtained microemulsions. Micelles made of cationic surfactants were a template for silica shell formation after adding TEOS [10].

3.2. Characterization of Silica Nanocapsules

SEM and STEM pictures of synthesised silica nanocapsules are presented in Figure 3. Obtained nanocontainers have a spherical shape with a diameter ranging between 100 nm and 200 nm. In STEM images (Figure 3c,d), a distinction (red arrow) between the wall and the core of the nanocapsules can be noticed, which confirms that obtained nanomaterials are capsules.

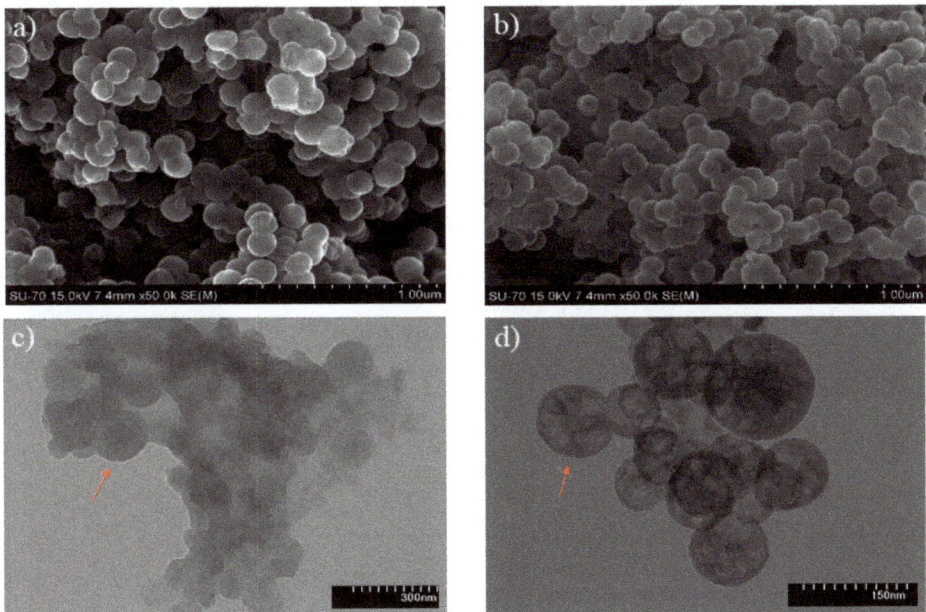

Figure 3. SEM (**a,b**) and scanning transmission electron microscopy (STEM) (**c,d**) images of silica nanocapsules based on N-hexadecyl-N,N,N-trimethylammonium bromide (CTAB) (**a,c**) and QSB2-12 (**c,d**).

In the FTIR spectra, presented in Figure 4a–c, some characteristic bands for silica capsules can be observed, such as Si–O–Si stretching at 1049 cm^{-1}, Si–OH stretching at 935 cm^{-1} and Si–O–Si bending at 800 cm^{-1}. By overlapping the spectra of SiNC_CTAB and SiNC_QSB2-12 (Figure 4c) it is possible to observe their similarity. FTIR spectra for CTAB and QSB2-12 show characteristic wavenumbers values of the bonds: C-H stretching absorption vibrations of long alkyl chains between 2840 cm^{-1} to 3000 cm^{-1}, C-H bending vibration of the $(CH_3)_4N^+$ cation at around 1490 cm^{-1} and rocking vibrations of $(CH_2)_n$, when n ≥ 4, at around 720 cm^{-1} (present only in case of long alkyl chains) [25]. In the silica capsules' spectra, lower intensity of signals associated with CTAB and QSB2-12 are observed suggesting that traces of cationic surfactants remained on the capsules' surface, as can be seen in the overlapped spectra of silica nanocapsules and surfactants (Figure 4a,b). Washing silica nanocapsules with an ethanolic solution of hydrochloric acid (1.5 mL of HCl in 150 mL of ethanol) at 60 °C allows the removal of cationic surfactants completely [16]. This route can be used for empty nanomaterials but applying it to nanocapsules loaded with active molecules may cause the unwanted release of the active molecules during washing steps. Another possibility to remove traces of cationic surfactants is calcination. However, this process removes all the organic compounds and cannot be applied to nanomaterials loaded with organic molecules [9,11].

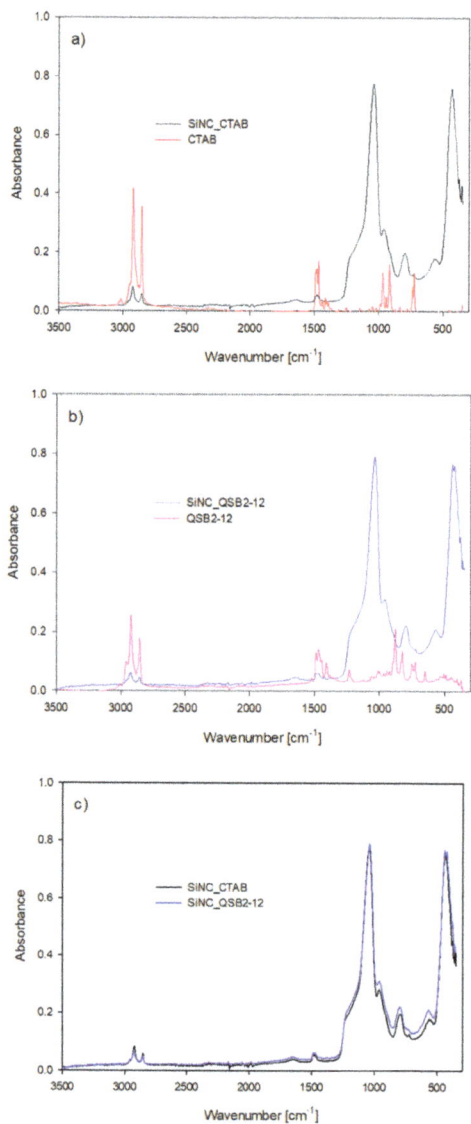

Figure 4. Overlapped Fourier-transformed infrared spectroscopy (FTIR) spectra of (**a**) SiNC_CTAB and CTAB, (**b**) SiNC_QSB2-12 and QSB2-12, and (**c**) SiNC_CTAB nd SiNC_QSB2-12.

Regarding the size distribution of the synthesised capsules in distilled water (Table 1), determined by DLS, peaks centred at 157.1 nm and 191.6 nm were observed for SiNC_CTAB and SiNC_QSB2-12, respectively (Figure 5). These data agree with the size determined by SEM, although polydispersion index (PdI) shows values higher than 0.5, indicating heterogeneity of the samples and the presence of larger aggregates. For both SiNC_CTAB and SiNC_QSB2-12 larger particles are observed probably due to the presence of polymer residues (TEOS polymerization) [14]. DLS data also confirm this assumption by the presence of secondary peaks, indicating the agglomeration of the nanocapsules and the polymer residues into larger particles [14]. The zeta potential associated with SiO_2 is typically −28 mV [31], so the recorded positive values of Z-potential for the synthesised silica nanocapsules in this work (cf. Table 1) are associated with the traces of cationic surfactants remaining in the capsules, which is consistent with FTIR spectra. Data for SiNC_CTAB are also in agreement with the literature where the hydrodynamic size for that conventional nanomaterial range from 80 nm to 180 nm [10,11,14].

Table 1. Dynamic Light Scattering (DLS) data for synthesized SiNC_CTAB and SiNC_QSB2-12.

	Hydrodynamic Size [nm]	ζ [mV]	PdI
SiNC_CTAB	157.1 ± 18.3 731.2 ± 134.3	32.8 ± 1.1	0.68
SiNC_QSB2-12	191.6 ± 37.1 662.1 ± 160.0	40.9 ± 6.8	0.56

ζ—Zeta potential; PdI—Polydispersion index.

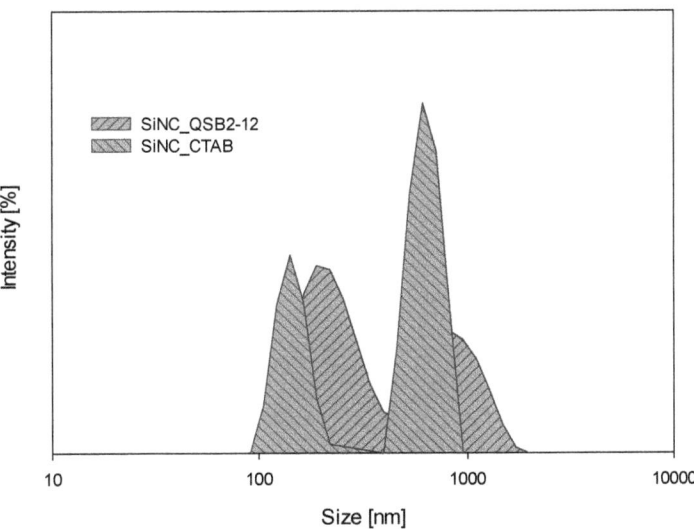

Figure 5. Size distribution of silica nanocontainers (SiNC) synthesized with CTAB and QSB2-12.

Thermogravimetric (TG) experiments were performed in order to verify and compare the thermal stability of synthesized nanocapsules. Tests with calcined samples (after thermal treatment) were carried out as a reference. TG profiles are presented in Figure 6. The degradation temperature for both SiNC_CTAB and SiNC_QSB2-12 was found to be 105 °C, which can be assigned to the process of dehydration of water adsorbed in the interlayer structure [32]. As expected for inorganic silica materials, curves for calcined nanomaterials show no variation in temperature, which suggests good thermal stability. Comparing curves for calcined and as-synthesized nanocapsules, a mass loss of approximately 80% can be observed for the as-synthesized ones. This may be due to the degradation of non-hydrolysed TEOS as well as some residues of cationic surfactants used for the synthesis [14]. These results suggest that replacing CTAB with QSB2-12 will not change the thermogravimetric profile of silica nanocapsules.

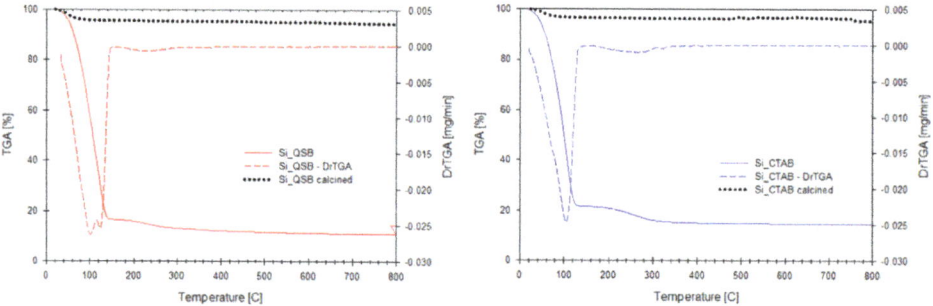

Figure 6. TG profile of SiNC_CTAB and SiNC_QSB2-12.

Textural properties of silica nanocontainers (SiNC_CTAB and SiNC_QSB2-12) were evaluated by adsorption–desorption isotherms N2 at −196 °C. In Figure 7a it is possible to observe that both samples show a typical type IV adsorption–desorption, according to IUPAC classification, with symmetric adsorption–desorption pathway isotherm indicating the presence mesoporous materials. A characteristic type H3 hysteresis loop is observed. This type is typical for mesoporous materials with slit-shaped pores [33]. Table 2 presents the main textural parameters of the silica nanocontainers: specific surface area (SBET) and the most frequent diameter of pores (mode of distribution of pore diameter). Data obtained for SiNC_CTAB are in agreement with what is reported in the literature when using diethyl ether as co-solvent [9]. When CTAB is replaced with QSB2-12 a slight increase in pore size was observed. Additionally, changing surfactants during synthesis results in a decrease in surface area from 719 m^2/g to 603 m^2/g.

Table 2. Textural properties of SiNC_CTAB and SiNC_QSB2-12.

	S$_{BET}$ [m^2/g]	Pore Size [nm]
SiNC_CTAB	719	4.4
SiNC_QSB2-12	603	5.9

3.3. Ecotoxicity

Ecotoxicity data (I/LC$_{50}$) are summarized in Table 3. Respective dose–response curves supporting these results are presented in a Supplementary Material (Figure S1). Both nanomaterials were toxic towards the tested microalgae and harmful towards the tested crustaceans [30]. However, SiNC_CTAB was more toxic than the novel SiNC_QSB2-12, for all tested species, statistically significant ($p < 0.05$) in the case of the diatom *P. tricornutum* (up to 3.5-times more toxic) and *T. chuii* (+25%) (Table 3). It has been suggested that the toxicity of silica nanocapsules may be associated with traces of cationic surfactants used during synthesis [10]. This is in accordance with FTIR data which confirm the presence of cationic surfactant residues in both nanomaterials. Previous studies by researchers from the University of Aveiro have shown that CTAB has 20-times higher toxicity towards *P. tricornutum* than gemini surfactant QSB2-12 [25]. These results prompted the decision to replace CTAB with QBS12, leading to the development of the work presented in this paper.

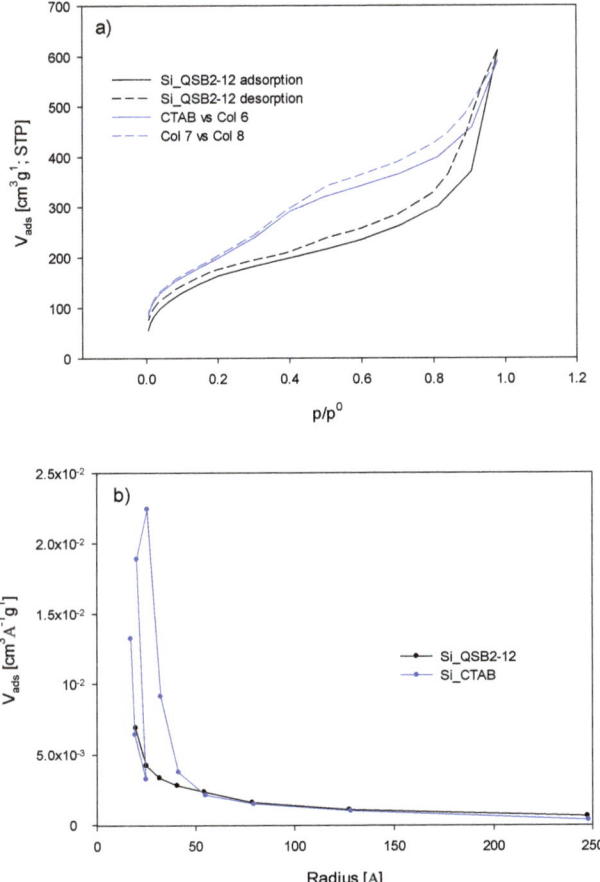

Figure 7. (**a**) Nitrogen adsorption–desorption isotherms and (**b**) pore size distribution of silica nanocapsules.

Table 3. Toxicity of synthesized reference (SiNC_CTAB) and newly developed (SiNC_QSB2-12) silica nanocapsules towards marine species and statistical comparison of I/LC$_{50}$ through t-tests and F-tests per tested species.

Marine Species	I/LC$_{50}$ [mg/L] (95% CI)		t-Test	p-Value	F-Test to Compare Variances DFn, DFd	p-Value
	SiNC_CTAB	SiNC_QSB2-12				
Nannochloropsis gaditana	8.51 (6.76–10.7)	9.69 (9.48–9.91)	1.2230	0.2282	94.91 20, 22	<0.0001
Tetraselmis chuii	7.41 (7.37–7.45)	9.26 (8.97–9.55)	3.3437	0.0014	17.28 19, 22	<0.0001
Phaeodactylum tricornutum	2.37 (1.75–3.23)	8.30 (2.69–15.61)	2.1350	0.0386	14.84 22, 20	<0.0001
Artemia salina	23.0 (20.6–25.7)	25.4 (18.4–35.0)	0.6126	0.5451	8.42 14, 14	0.0003

Data are presented as median growth inhibition concentration (72 h-IC$_{50}$) and median lethal concentration values (48 h-LC$_{50}$), and respective 95% confidence intervals (CI). DFn—degrees of freedom numerator. DFd—degrees of freedom denominator.

There are some ecotoxicity data available in the literature for CTAB-based silica nanocapsules. For the diatom *P. tricornutum* IC_{50} ranges from 2.03 mg/L [10] to 3.67 mg/L [13], for microalgae *N. gaditana* IC_{50} is 1.15 mg/L [10], while for the crustacean *A. salina* SiNC_CTAB are harmful (LC_{50} = 12.2 mg/L [34]) or non-toxic (LC_{50} > 100 mg/L [10]). The results obtained in the present work agree with the literature data, particularly in the case of diatoms and crustaceans. There is a difference for the microalgae *N. gaditana*, which may be associated with different amounts of the cationic surfactant still adsorbed to the silica nanocapsules surface.

Sustainability is, nowadays, one of the main challenges in materials science. Designing nanomaterials by a proper selection of raw materials seems to be the way to obtain environmentally-friendly nanocontainers. Since it has been reported that the encapsulation of active molecules reduces their toxicity due to a controlled release mechanism [10], non-toxic nanocarriers are needed. To the best of our knowledge, there is no information published on the use of different cationic surfactants in order to prepare silica nanocapsules. However, comparing data of the toxicity profile of SiNC_CTAB with another nanomaterial namely, layered double hydroxides (LDHs) the silica nanocarriers have been found as exerting some degree of toxicity towards marine species [12]. Other eco-friendly alternatives have been reported in the literature, such as natural polymercarriers based on gelatine or chitosan microcapsules. These materials are non-toxic, biocompatible and biodegradable [35,36]. However, due to some limitations, such as size or swelling in aqueous solutions, their use is not always possible. In those cases, silica nanocapsules are often a solution, however current versions are still toxic.

This study demonstrates that developing greener silica mesoporous nanocapsules by replacing cationic surfactants used during SiNC synthesis, may lead to less toxic alternatives. In the present study, this was not fully achieved, however, the statistical differences between the toxicity on two ecologically relevant marine species demonstrate this approach is in the right direction to obtain less toxic materials. Considering the potential application of such nanomaterials as coating additives for maritime applications, improvements in their design and further ecotoxicological studies on other marine species may confirm the present findings.

4. Conclusions

Silica nanocapsules based on gemini surfactant QSB2-12 were successfully synthesised and characterized.

The synthesis is performed in a one-stage process where the toxic commercial cationic surfactant CTAB is replaced by a less toxic gemini surfactant QSB2-12. Changing surfactants allows the reduction, by 50%, of the number of moles used to obtain a stable microemulsion and does not hinder the formation of the nanomaterials. Moreover, the present results show that replacing the surfactant does not significantly change the structure and properties of silica nanocapsules.

Ecotoxicity tests show that nanocapsules based on CTAB are significantly more toxic than nanocontainers prepared with gemini surfactant for two (out of four) species and slightly more toxic for the other two. Both nanomaterials are classified as toxic (microalgae) and harmful (crustacean). The present study shows that by choosing a template surfactant carefully, environmentally safer silica nanocapsules can be designed. Nonetheless, further research is still needed.

To the best of our knowledge, the safe-by-design approach for silica nanocapsules synthesis is addressed in this work for the first time. Following this approach allows us to obtain not only environmentally friendly nanomaterials but also prevent the use of toxic surfactants and decrease the number of used chemicals, thus being aligned with the concept of green chemistry [37].

Supplementary Materials: The following are available online at http://www.mdpi.com/2076-3417/10/22/8085/s1, Figure S1. Dose–response curves of the microalgae *Nannochloropsis gaditana* and *Tetraselmis chuii*, the diatom *Phaeodactylum tricornutum* and the crustacean *Artemia salina* exposed to tested cationic surfactants.

Author Contributions: Conceptualization, O.K., R.M. and J.T.; Methodology, O.K., R.M. and I.S.; Investigation, O.K., I.S. and J.F.; Resources, O.K. and R.M.; Data Curation, O.K. and J.F.; Writing—Original Draft, O.K.; Writing—Review & Editing, I.S., R.M., S.L. and J.T.; Visualization, O.K. and J.F.; Project administration, O.K.;

Funding acquisition, O.K.; Validation, R.M.; Supervision: R.M., S.L. and J.T. All authors have read and agreed to the published version of the manuscript.

Funding: O. Kaczerewska received funding from the European Union's Horizon 2020 research and innovation programme under the Marie Sklodowska-Curie grant agreement No 792945 (EcoGemCoat). R. Martins was hired under the Scientific Employment Stimulus—Individual Call (CEECIND/01329/2017), funded by national funds (OE), through FCT, in the scope of the framework contract foreseen in the numbers 4, 5 and 6 of the article 23, of the Decree-Law 57/2016, of 29 August, changed by Law 57/2017, of 19 July. This work was developed under the framework of the NANOGREEN project (CIRCNA/BRB/0291/2019), funded by the Portuguese Foundation for Science and Technology (FCT) through national funds (OE). This work was also carried out in the framework of SMARTAQUA project, which is funded by the Foundation for Science and Technology in Portugal (FCT), the Research Council of Norway (RCN-284002), Malta Council for Science and Technology (MCST), and co-funded by European Union's Horizon 2020 research and innovation program under the framework of ERA-NET Cofund MarTERA (Maritime and Marine Technologies for a new Era). This work was developed within the scope of the project CICECO-Aveiro Institute of Materials (UIDB/50011/2020; UIDP/50011/2020) and CESAM–Centre for Environmental and Marine Studies (UIDB/50017/2020 + UIDP/50017/2020) financed by national funds.

Conflicts of Interest: The authors declare no conflict of interest.

References

1. Mirmohseni, A.; Akbari, M.; Najjar, R.; Hosseini, M. Self-healing waterborne polyurethane coating by pH-dependent triggered-release mechanism. *J. Appl. Polym. Sci.* **2019**, *136*, 47082. [CrossRef]
2. Zhao, D.; Liu, D.; Hu, Z. A smart anticorrosion coating based on hollow silica nanocapsules with inorganic salt in shells. *J. Coat. Technol. Res.* **2017**, *14*, 85–94. [CrossRef]
3. Galvão, T.L.P.; Sousa, I.; Wilhelm, M.; Carneiro, J.; Opršal, J.; Kukačková, H.; Špaček, V.; Maia, F.; Gomes, J.R.B.; Tedim, J.; et al. Improving the functionality and performance of AA2024 corrosion sensing coatings with nanocontainers. *Chem. Eng. J.* **2018**, *341*, 526–538. [CrossRef]
4. Zheludkevich, M.L.; Tedim, J.; Ferreira, M.G.S. "Smart" coatings for active corrosion protection based on multi-functional micro and nanocontainers. *Electrochim. Acta* **2012**, *82*, 314–323. [CrossRef]
5. Wang, J.-X.; Wang, Z.-H.; Chen, J.-F.; Yun, J. Direct encapsulation of water-soluble drug into silica microcapsules for sustained release applications. *Mater. Res. Bull.* **2008**, *43*, 3374–3381. [CrossRef]
6. Samadzadeh, M.; Boura, S.H.; Peikari, M.; Kasiriha, S.M.; Ashrafi, A. A review on self-healing coatings based on micro/nanocapsules. *Prog. Org. Coat.* **2010**, *68*, 159–164. [CrossRef]
7. Jain, A.K.; Thareja, S. In vitro and in vivo characterization of pharmaceutical nanocarriers used for drug delivery. *Artif. Cells Nanomed. Biotechnol.* **2019**, *47*, 524–539. [CrossRef] [PubMed]
8. Pathak, C.; Vaidya, F.U.; Pandey, S.M. Mechanism for Development of Nanobased Drug Delivery System. In *Applications of Targeted Nano Drugs and Delivery Systems*; Elsevier: Amsterdam, The Netherlands, 2019; pp. 35–67, ISBN 978-0-12-814029-1.
9. Chen, H.; He, J.; Tang, H.; Yan, C. Porous Silica Nanocapsules and Nanospheres: Dynamic Self-Assembly Synthesis and Application in Controlled Release. *Chem. Mater.* **2008**, *20*, 5894–5900. [CrossRef]
10. Figueiredo, J.; Oliveira, T.; Ferreira, V.; Sushkova, A.; Silva, S.; Carneiro, D.; Cardoso, D.; Goncalves, S.; Maia, F.; Rocha, C.; et al. Toxicity of innovative anti-fouling nano-based solutions in marine species. *Environ. Sci. Nano* **2019**, *6*, 1418–1429. [CrossRef]
11. Maia, F.; Silva, A.P.; Fernandes, S.; Cunha, A.; Almeida, A.; Tedim, J.; Zheludkevich, M.L.; Ferreira, M.G.S. Incorporation of biocides in nanocapsules for protective coatings used in maritime applications. *Chem. Eng. J.* **2015**, *270*, 150–157. [CrossRef]
12. Avelelas, F.; Martins, R.; Oliveira, T.; Maia, F.; Malheiro, E.; Soares, A.M.V.M.; Loureiro, S.; Tedim, J. Efficacy and Ecotoxicity of Novel Anti-Fouling Nanomaterials in Target and Non-Target Marine Species. *Mar. Biotechnol.* **2017**, *19*, 164–174. [CrossRef] [PubMed]
13. Gutner-Hoch, E.; Martins, R.; Oliveira, T.; Maia, F.; Soares, A.; Loureiro, S.; Piller, C.; Preiss, I.; Weis, M.; Larroze, S.; et al. Antimacrofouling Efficacy of Innovative Inorganic Nanomaterials Loaded with Booster Biocides. *J. Mar. Sci. Eng.* **2018**, *6*, 6. [CrossRef]
14. Maia, F.; Tedim, J.; Lisenkov, A.D.; Salak, A.N.; Zheludkevich, M.L.; Ferreira, M.G.S. Silica nanocontainers for active corrosion protection. *Nanoscale* **2012**, *4*, 1287–1298. [CrossRef] [PubMed]
15. Lang, N.; Tuel, A. A Fast and Efficient Ion-Exchange Procedure to Remove Surfactant Molecules from MCM-41 Materials. *Chem. Mater.* **2004**, *16*, 1961–1966. [CrossRef]

16. Hofmann, C.; Duerkop, A.; Baeumner, A.J. Nanocontainers for Analytical Applications. *Angew. Chem. Int. Ed.* **2019**, *58*, 12840–12860. [CrossRef]
17. Brycki, B.E.; Kowalczyk, I.H.; Szulc, A.; Kaczerewska, O.; Pakiet, M. Multifunctional Gemini Surfactants: Structure, Synthesis, Properties and Applications. In *Application and Characterization of Surfactants*; Najjar, R., Ed.; InTech: London, UK, 2017; ISBN 978-953-51-3325-4.
18. Garcia, M.T.; Kaczerewska, O.; Ribosa, I.; Brycki, B.; Materna, P.; Drgas, M. Hydrophilicity and flexibility of the spacer as critical parameters on the aggregation behavior of long alkyl chain cationic gemini surfactants in aqueous solution. *J. Mol. Liq.* **2017**, *230*, 453–460. [CrossRef]
19. Goronja, J.; Janosevic-Lezaic, A.; Dimitrijevic, B.; Malenovic, A.; Stanisavljev, D.; Pejic, N. Determination of critical micelle concentration of cetyltrimethylammonium bromide: Different procedures for analysis of experimental data. *Hem. Ind.* **2016**, *70*, 485–492. [CrossRef]
20. Garcia, M.T.; Kaczerewska, O.; Ribosa, I.; Brycki, B.; Materna, P.; Drgas, M. Biodegradability and aquatic toxicity of quaternary ammonium-based gemini surfactants: Effect of the spacer on their ecological properties. *Chemosphere* **2016**, *154*, 155–160. [CrossRef]
21. European Chemicals Agency. *Cetrimonium Bromide*—Registration Dossier (EC Number: 200-311-3). 2020. Available online: https://echa.europa.eu/brief-profile/-/briefprofile/100.000.283 (accessed on 11 November 2020).
22. Garcia, M.T.; Ribosa, I.; Kowalczyk, I.; Pakiet, M.; Brycki, B. Biodegradability and aquatic toxicity of new cleavable betainate cationic oligomeric surfactants. *J. Hazard. Mater.* **2019**, *371*, 108–114. [CrossRef]
23. Zhang, C.; Cui, F.; Zeng, G.; Jiang, M.; Yang, Z.; Yu, Z.; Zhu, M.; Shen, L. Quaternary ammonium compounds (QACs): A review on occurrence, fate and toxicity in the environment. *Sci. Total Environ.* **2015**, *518–519*, 352–362. [CrossRef]
24. European Chemicals Agency. *Dodecyltrimethylammonium Chloride*—Registration Dossier (EC Number: 203-927-0). 2020. Available online: https://echa.europa.eu/substance-information/-/substanceinfo/100.003.570 (accessed on 11 November 2020).
25. Kaczerewska, O.; Martins, R.; Figueiredo, J.; Loureiro, S.; Tedim, J. Environmental behaviour and ecotoxicity of cationic surfactants towards marine organisms. *J. Hazard. Mater.* **2020**, *392*, 122299. [CrossRef] [PubMed]
26. Van de Poel, I.; Robaey, Z. Safe-by-Design: From Safety to Responsibility. *Nanoethics* **2017**, *11*, 297–306. [CrossRef] [PubMed]
27. Nymark, P.; Bakker, M.; Dekkers, S.; Franken, R.; Fransman, W.; García-Bilbao, A.; Greco, D.; Gulumian, M.; Hadrup, N.; Halappanavar, S.; et al. Toward Rigorous Materials Production: New Approach Methodologies Have Extensive Potential to Improve Current Safety Assessment Practices. *Small* **2020**, *16*, 1904749. [CrossRef] [PubMed]
28. OECD Test No. 201: Freshwater alga and cyanobacteria, growth inhibition test. In *OECD Guidelines for the Testing of Chemicals*; Organization for Economic Co-Operation and Development: Paris, France, 2011; ISBN 978-92-64-06992-3.
29. ISO. *ISO 10253 Water Quality—Marine Algal Growth Inhibition Test with Skeletonema sp. and Phaeodactylum tricornutum*; International Organization for Standardization: Geneva, Switzerland, 2016.
30. Blaise, C.; Gagné, F.; Férard, J.F.; Eullaffroy, P. Ecotoxicity of selected nano-materials to aquatic organisms. *Environ. Toxicol.* **2008**, *23*, 591–598. [CrossRef]
31. Sousa, I.; Maia, F.; Silva, A.; Cunha, A.; Almeida, A.; Evtyugin, D.V.; Tedim, J.; Ferreira, M.G. A novel approach for immobilization of polyhexamethylene biguanide within silica capsules. *RSC Adv.* **2015**, *5*, 92656–92663. [CrossRef]
32. Kozak, M.; Domka, L. Adsorption of the quaternary ammonium salts on montmorillonite. *J. Phys. Chem. Solids* **2004**, *65*, 441–445. [CrossRef]
33. Alothman, Z. A Review: Fundamental Aspects of Silicate Mesoporous Materials. *Materials* **2012**, *5*, 2874–2902. [CrossRef]
34. Gutner-Hoch, E.; Martins, R.; Maia, F.; Oliveira, T.; Shpigel, M.; Weis, M.; Tedim, J.; Benayahu, Y. Toxicity of engineered micro- and nanomaterials with antifouling properties to the brine shrimp Artemia salina and embryonic stages of the sea urchin Paracentrotus lividus. *Environ. Pollut.* **2019**, *251*, 530–537. [CrossRef]
35. Hiwale, P.; Lampis, S.; Conti, G.; Caddeo, C.; Murgia, S.; Fadda, A.M.; Monduzzi, M. In Vitro Release of Lysozyme from Gelatin Microspheres: Effect of Cross-linking Agents and Thermoreversible Gel as Suspending Medium. *Biomacromolecules* **2011**, *12*, 3186–3193. [CrossRef]

36. Sousa, I.; Quevedo, M.C.; Sushkova, A.; Ferreira, M.G.S.; Tedim, J. Chitosan Microspheres as Carriers for pH-Indicating Species in Corrosion Sensing. *Macromol. Mater. Eng.* **2019**, 1900662. [CrossRef]
37. Saleh, H.E.-D.M.; Koller, M. Introductory Chapter: Principles of Green Chemistry. In *Green Chemistry*; Saleh, H.E.-D.M., Koller, M., Eds.; IntechOpen: London, UK, 2018.

Publisher's Note: MDPI stays neutral with regard to jurisdictional claims in published maps and institutional affiliations.

© 2020 by the authors. Licensee MDPI, Basel, Switzerland. This article is an open access article distributed under the terms and conditions of the Creative Commons Attribution (CC BY) license (http://creativecommons.org/licenses/by/4.0/).

Article

Can Encapsulation of the Biocide DCOIT Affect the Anti-Fouling Efficacy and Toxicity on Tropical Bivalves?

Juliana Vitoria Nicolau dos Santos [1], Roberto Martins [2,*], Mayana Karoline Fontes [1], Bruno Galvão de Campos [1], Mariana Bruni Marques do Prado e Silva [1], Frederico Maia [3], Denis Moledo de Souza Abessa [1] and Fernando Cesar Perina [1]

1. NEPEA—Núcleo de Estudos sobre Poluição e Ecotoxicologia Aquática, Universidade Estadual Paulista, Campus do Litoral Paulista (UNESP-CLP), São Vicente 11330-900, Brazil; juliana_juv@hotmail.com (J.V.N.d.S.); mayanakf@gmail.com (M.K.F.); brunog_campos@yahoo.com.br (B.G.d.C.); marianabruni.prado@gmail.com (M.B.M.d.P.e.S.); denis.abessa@unesp.br (D.M.d.S.A.); perinafc@gmail.com (F.C.P.)
2. CESAM—Centre for Environmental and Marine Studies and Department of Biology, University of Aveiro, Campus Universitário de Santiago, 3810-193 Aveiro, Portugal
3. Smallmatek—Small Materials and Technologies, Lda., Rua Canhas, 3810-075 Aveiro, Portugal; frederico.maia@smallmatek.pt
* Correspondence: roberto@ua.pt; Tel.: +351-234-370-317

Received: 4 November 2020; Accepted: 27 November 2020; Published: 30 November 2020

Featured Application: The present study can foster nanoecotoxicology research on tropical environments and can be also applied on the maritime industry.

Abstract: The encapsulation of the biocide DCOIT in mesoporous silica nanocapsules (SiNC) has been applied to reduce the leaching rate and the associated environmental impacts of coatings containing this biocide. This research aimed to evaluate the effects of DCOIT in both free and nanostructured forms (DCOIT vs. SiNC-DCOIT, respectively) and the unloaded SiNC on different life stages of the bivalve *Perna perna*: (a) gametes (fertilization success), (b) embryos (larval development), and (c) juveniles mussels (byssus threads production and air survival after 72 h of aqueous exposure). The effects on fertilization success showed high toxicity of DCOIT (40 min-EC_{50} = 0.063 µg L^{-1}), followed by SiNC-DCOIT (8.6 µg L^{-1}) and SiNC (161 µg L^{-1}). The estimated 48 h-EC_{50} of SiNC, DCOIT and SiNC-DCOIT on larval development were 39.8, 12.4 and 6.8 µg L^{-1}, respectively. The estimated 72 h-EC_{50} for byssus thread production were 96.1 and 305.5 µg L^{-1}, for free DCOIT and SiNC-DCOIT, respectively. Air survival was significantly reduced only for mussels exposed to free DCOIT. Compared to its free form, SiNC-DCOIT presented a balanced alternative between efficacy and toxicity, inhibiting efficiently the development of the target stage (larvae that is prone to settle) and satisfactorily preventing the juvenile attachment.

Keywords: *Perna perna*; biofouling; nanotechnology; toxicity

1. Introduction

Long-term submerged structures are susceptible to aggregate fouling organisms, such as algae, barnacles, mussels, and other benthic organisms, known as biofouling [1]. In the shipping industry, this phenomenon causes extensive economic losses, resulting in a significant decrease in ships' durability and operational efficiency, interfering with vessels' navigability. The friction caused by the increase in the hull's roughness exponentially increases maritime transport costs, as it demands greater engine power and fuel consumption [2]. Moreover, this increase in fuel consumption leads to an increase in greenhouse gas emissions. Another problem related with the biofouling regards the dispersion of invasive species associated with the hull's vessels or its ballast water [3].

However, the traditional methods used to inhibit the establishment and growth of biofouling have arisen concerns since the 20th century due to the critical environmental impacts, particularly on non-target foulers/biota. The chemical formulation of antifouling (AF) paints contain biocides, which are chemical substances that neutralize, inhibit or exert control over undesirable fouling organisms and/or communities, preventing their settlement and further growth [4]. Despite the desired antifouling action, many of these compounds are toxic to non-target species [5,6], and may cause adverse effects to the non-fouling biota. The global ban on the organotin-based paints in 2008 increased the use of alternative AF booster biocides, such as 4,5-dichloro-2-octyl-2*H*-isothiazole-3-one (DCOIT). DCOIT is the biocidal ingredient of the AF products Sea Nine 211™ or Kathlon™ 910 SB, among other commercial products, which stands out as one of the most widely used AF agents in maritime topcoats [7]. This organic compound has been considered environmentally safe by USEPA [8], due to its reduced half-life (<1 day in seawater) [9,10]. However, recent studies have indicated that its half-life can be longer than four days in natural seawater [11], at least one week in artificial saltwater [12], and up to 13 days [13], depending on the environmental conditions, such as sunlight, dissolved oxygen or temperature [1]. Consequently, DCOIT has been found in water and sediment from several countries in Asia [14] and Europe [15]. Not surprisingly, studies have demonstrated this biocide's high toxicity to non-target organisms [12,16]. Recently, DCOIT was classified as "very toxic to aquatic life, with long-lasting effects" by the European Chemicals Agency [17].

Because environmental regulations have become increasingly restrictive on national and international bases, new AF strategies need to attend some aspects such as environmental safety, AF efficacy and coating's lifetime. In this sense, promising AF alternatives to regular state-of-the-art biocides have gained space, such as the encapsulation and immobilization of the biocides in low toxic nanocontainers [18–22]. Novel AF nanomaterials have been recently developed using silica mesoporous nanocapsules (SiNC) to encapsulate DCOIT [23]. SiNC have an empty core and shell with gradual mesoporosity which confers significant loading capacity and allows prolonged and stimuli-triggered release of the biocide, such as pH or chloride concentration [24], providing a safe environmental loading of the biocide together with specific targeting [23]. When used as coating additive, the biocide's lifetime is increased and the percentage content of the main active ingredient in the coating is also greatly reduced. This feature has been triggering the core business of the company Smallmatek, Lda, which develops and produces functionalized engineered nanomaterials, including SiNC-DCOIT, to be added in protective coatings for maritime applications.

A recent study comparing the antifouling efficacy and toxicity of DCOIT in both free and encapsulated forms (DCOIT vs. SiNC-DCOIT) indicated that the novel AF nanostructured additive is much less toxic (up to 214-fold) towards non-target species from temperate regions [12]. However, there is no knowledge on the toxicity and efficacy of this AF nanomaterial on species from different climatic regions, a critical requirement on environmental risk assessment at global scale. Furthermore, the effects of SiNC-DCOIT on different life stages of a given species are also unknown so far. The brown mussel *Perna perna* occurs on rocky reefs, forming dense colonies at the low-tidal and intertidal levels in the tropical and subtropical zones. This bivalve species is regarded as a socio-economically relevant species with a very-broad natural distribution along Africa and the Arabic Peninsula, being an invasive species in the Gulf of Mexico and, more recently, in Portugal [25]. This bivalve species is economically important in Brazilian southern and southeastern coastal areas and represents a relevant food resource for coastal populations [26]. Therefore, due to its socio-economic importance, *P. perna* is a non-target fouling organism at adult life stages; however, it can be considered a target organism at the larval stage since it is prone to settle and become part of the undesirable fouling in human-made structures. Accordingly, this research aimed to evaluate and compare the effect of two DCOIT forms (free DCOIT vs. its nanostructured form SiNC-DCOIT) and the unloaded nanocarriers (SiNC) on different life stages of the bivalve *P. perna*, namely, on gametes (fertilization success), fertilized eggs (embryo-larval development), and juveniles mussels (byssus threads production and air survival capacity).

2. Materials and Methods

2.1. Chemical Compounds

DCOIT (CAS nr. 64359-81-5) was purchased from Sigma-Aldrich (São Paulo, SP, Brazil). Nanomaterials (SiNC; SiNC-DCOIT) were supplied by Smallmatek, Lda. (Aveiro, Portugal). Tested nanomaterials were fully characterized by Figueiredo et al. [12]. Briefly, SiNC has a diameter of 129 nm and SiNC-DCOIT has diameter of 152 nm and a biocidal content of 18.3% [12]. Stock solutions/dispersions were prepared in natural seawater (salinity 33 ± 2), filtered through a 0.22 μm microporous membrane filter and dispersed in an ultrasonic bath (40 kHz) for 30 min. Diluted dispersions were sonicated for 15 min., immediately before the exposure test.

2.2. Animals

Adult and juvenile *Perna perna* mussels were acquired from a mariculture farm located on the north coast of São Paulo (Brazil). In the laboratory, the animals were placed on 60 L plastic boxes filled with seawater (salinity 33) and maintained under constant aeration, temperature (25 ± 2 °C), and photoperiod of 12 h:12 h (light:dark) for 72 h prior to the experiments [27].

2.3. Fertilization Assay

The fertilization assay was carried out according to the protocol proposed by the United States Environmental Protection Agency [28], adapted for *P. perna* by Zaroni et al. [27]. The exposure concentrations used in the fertilization assay were: 1, 3.33, 10, 33 and 100 μg L^{-1} of free DCOIT; 24.7, 74.1, 222.2, 666.7 and 2000 μg L^{-1} for SiNC-DCOIT (expressed as DCOIT content); and 250, 500, 1000, 2000 and 4000 μg L^{-1} for SiNC. These concentration ranges were chosen based on previous toxicity data acquired for temperate marine species [12].

The mussel's gametes (eggs and sperm) were obtained by the thermal induction [27]. The experiment was carried out in glass test tubes containing 10 mL of each test solution. Four replicates were prepared for each treatment and experimental control (filtered seawater). In each replicate, 150 μL of sperm solution were transferred to each replicate and incubated (25 ± 2 °C) for 40 min. Next, approximately 2000 eggs were added to the sperm suspensions. Sixty minutes after the eggs are added, the test was terminated by adding 500 μL of formaldehyde (10%) buffered with borax to each replicate. The percent fertilization was determined by microscopic examination of an aliquot from each replicate of the treatments placed on a Sedgwick-Rafter chamber. The first 100 eggs observed were counted and the average percentage of fertilization was determined, considering the 4 replicates used.

2.4. Embryo-Larval Development Assay

The short-term chronic exposure assay using *P. perna* embryos followed the standard method ABNT NBR 16,456 [29]. In this experiment, six exposure concentrations were tested for each chemical: 1, 3, 3.3, 10, 33 and 100 μg L^{-1} of free DCOIT, 0.064, 0.32, 1.6, 8, 40 and 200 μg (of DCOIT) L^{-1} in SiNC-DCOIT, and 6.5, 32, 162, 808 and 4040 μg L^{-1} of SiNC. These concentrations were established based on the findings of a preliminary experiment.

Gametes obtention and fertilization were conducted according to the procedure described in the fertilization assay. Approximately 400–500 fertilized eggs were added to the glass test-tubes containing 10 mL of test solutions. Four replicates per concentration were prepared. The experiment was incubated (25 ± 2 °C) for 48 h with a photoperiod of 12 h:12 h (light:dark). Development confirmation was verified in the experimental control (above 70% of normal larvae, according to Zaroni et al., [27]). The test was then terminated by adding 500 μL of 10% buffered formaldehyde to each replicate. The first

100 organisms observed were counted and identified: and the normal veliger larvae were determined as those presenting symmetrical and closed valves, visible internal content, and a "D shape". On the other hand, abnormal larvae included those undeveloped, those exhibiting delays and/or morphological abnomalies and the absence of development of eggs.

2.5. Short-Term Exposure Assay with Juveniles

The tested concentrations were 10, 100 and 1000 µg L^{-1} for SiNC and SiNC-DCOIT (as mg of DCOIT L^{-1}) and 0.81, 8.1 and 81 µg L^{-1} for DCOIT (free form), based on the sublethal toxicity of these compounds for adult mussels of the species *Mytilus galloprovincialis* [12]. Negative control (filtered seawater) was also prepared.

Juvenile mussels with approximately 2.5 cm long were selected. Their byssus threads were cut with surgical scissors aiming at to examine their growth during exposure and counting at the end of the experiment. A total of 300 animals were randomly divided into 500 mL glass flasks with 400 mL of test solution ($n = 6$, 5 organisms per replicate). The animals were exposed for 72 h with constant gentle aeration, at 25 ± 2 °C and photoperiod of 12:12 h (light:dark). During the experimental period, mussels were not fed and the test solutions were not renewed. In order to maintain the quality of the experiment, survival of the organisms were evaluated daily and dead organisms were removed; physical-chemical parameters were measured in the beginning and in the end of the exposure.

2.5.1. Byssus Threads Formation

The number of byssus threads produced by the organisms ($n = 6$ per treatment) was counted after the exposure period. This procedure consisted of viewing both upside and underside of the mussel through the transparent glass flasks that were clearly visible, allowing the identification and counting of all the byssus threads.

2.5.2. Survival-in-Air Test

Mussels ($n = 3$) of each treatment were exposed to air, on empty containers (6-wells cell culture plates), after the aqueous exposure period of 72 h aiming at assessing the fitness to survive in prolonged extreme conditions. Containers were kept at constant room temperature (25 ± 1 °C) and photoperiod (12:12 h). Organisms' survival was checked daily by an inspection of the valve closure. Mussels were considered dead when their valves did not close after mechanical stimulation.

2.6. Statistical Analysis

Data normality and homoscedasticity obtained in each experiment (fertilization, embryotoxicity, byssus threads produced, and survival in air tests) were tested using the Shapiro Wilk and the Levene tests ($p < 0.05$), respectively. For each experiment, statistical differences between the negative control and each treatment were analyzed using one-way ANOVA, followed by Dunnett's multiple comparison tests whenever significant differences were observed ($p < 0.05$). Then, the no observed effect concentration (NOEC) and the lowest observed effect concentration (LOEC) were derived for fertilization and embryotoxicity tests.

The median effective concentrations (EC_{50}) that reduce 50% of fertilization success, embryo-larval development and byssus threads production were determined using dose-response curves through nonlinear regression analysis using a 4-parameter log-logistic model, performed with the statistical software GraphPad Prism v.6 (GraphPad Software, La Jolla, CA, USA). The number of byssus threads for each chemical concentration were normalized to the total threads production of the control (considered as 100%). For each chemical, the nonlinear regression equation that best fits the data was selected considering the R^2 value, absolute sum of squares, and the 95% confidence intervals.

3. Results and Discussion

3.1. Fertilization Assay

All tested treatments exhibited significant effects compared to the negative control (Figure 1; Table 1). The 40 min-EC_{50} value of DCOIT (free form) was estimated at 0.063 µg L^{-1} (Table 1), roughly 137-fold more toxic than the nanostructured form SiNC-DCOIT (8.6 µg DCOIT L^{-1}) and 4 orders of magnitude more toxic than the unloaded SiNC (161 µg L^{-1}). The high difference between toxicity of SiNC-DCOIT in comparison with DCOIT may be related to the biocidal controlled release property of SiNC-DCOIT, which occurs gradually in time and by predefined stimuli [12,13].

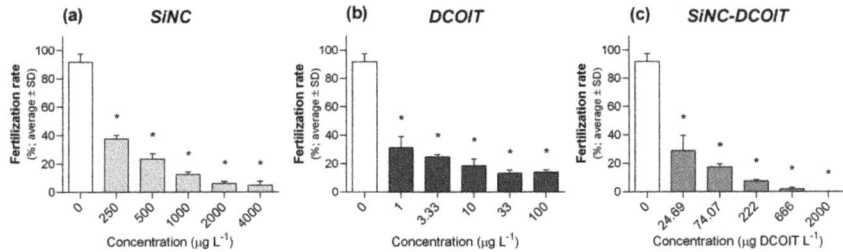

Figure 1. Fertilization rates observed for *P. perna* gametes exposed to (**a**) SiNC, (**b**) DCOIT, and (**c**) SiNC-DCOIT. Asterisks (*) indicate significant differences relative to the control ($p < 0.05$). Data are presented as average ± standard deviation (SD).

Table 1. NOEC, LOEC, EC_{50} values and respective confidence intervals (95%) of DCOIT, SiNC-DCOIT, and SiNC on gametes fertilization, embryo-larval development, byssus threads production, and survival-in-air of *P. perna*. Units are given in µg L^{-1}. SiNC-DCOIT values correspond to the concentration of encapsulated DCOIT (µg DCOIT L^{-1}). "nd": not determined.

	Parameter	NOEC	LOEC	EC_{50}	95% CI
SiNC	Fertilization	<250.0	250.0	161.3	139.3–186.8
	Embryotoxicity	<6.5	6.5	39.8	10.5–150.8
	Byssus threads	1000	>1000	1323	218.2–8019
	Air survival capacity	1000	>1000	nd	–
DCOIT	Fertilization	<1.0	1.0	0.063	0.016–0.255
	Embryotoxicity	1.0	3.3	12.4	9.9–15.4
	Byssus threads	81.0	>81.0	96.1	6.3–470
	Air survival capacity	<0.810	0.810	nd	–
SiNC-DCOIT	Fertilization	<24.7	24.7	8.6	4.9–15.0
	Embryotoxicity	0.064	0.320	6.8	2.7–16.9
	Byssus threads	100.0	1000	305.5	124.2–751.5
	Air survival capacity	<10.0	10.0	nd	–

These results stress the high toxicity of the conventional form of DCOIT by inhibiting the fertilization of *P. perna* gametes even at very low and environmentally relevant exposure concentrations. As a comparison, the EC_{50} reported for the sea urchin *Paracentrotus lividus* fertilization assay was 198 µg L^{-1} [30]. However, it is worth pointing out that the experiment was slightly different since sperm was pre-exposed for 45 min, and the subsequent fertilization was carried out in artificial seawater without DCOIT. In the present study, the fertilization was carried out in the same test-solutions that sperm was exposed for 40 min.

Previous studies demonstrate that DCOIT was more toxic (i.e., efficient) towards temperate target species, namely the bacterium *Vibrio fischeri* ($IC_{50} = 299$ µg L^{-1} of free DCOIT vs. $IC_{50} = 459$ µg DCOIT L^{-1} for SiNC-DCOIT) and the diatom *Phaeodactylum tricornutum*, involved in the biofilm formation

(IC_{50} = 4 μg L^{-1} and 7 μg L^{-1} for free DCOIT and SiNC-DCOIT, respectively [12]). Nevertheless, in such study, the sibling nanomaterial containing also silver (SiNC-DCOIT-Ag), demonstrated the opposite being much more efficient than the dissolved forms of DCOIT or Ag towards the tested target species [12].

3.2. Embryo-Larval Development Assay

The effects of SiNC, DCOIT, and SiNC-DCOIT on the larval development rate of *P. perna* are shown in Figure 2 and Table 1. Significant effects were observed as low as at the lowest tested exposure concentration of unloaded SiNC (6.5 μg L^{-1}). Additionally, at 4040 μg SiNC L^{-1} no developed veliger larvae were found (Figure 2a). The 48 h-EC_{50} value was set at 39.8 μg SiNC L^{-1}, the highest value amongst the three tested compounds. Despite being the less toxic among the tested chemicals, the embryotoxicity caused by SiNC was much higher than expected since silica was expected to be, in principle, inert. According to Figueiredo et al. [12] this effect can be attributed to the cationic surfactant cetyltrimethylammonium bromide (CTAB) used in the synthesis of the mesoporous silica nanocapsules [23]. CTAB is a quaternary ammonium compound (QACs), a class of toxic surfactants towards aquatic species [12,31]. Besides, residues of this organic component can be detected even after several washes [12] explaining the recently reported effects of SiNC on the fertilization and embryo-larval development of the sea-urchin *Paracentrotus lividus* [32] and the inhibition settlement of the bryozoan *Bugula neritina* larvae [33]. This raw nanomaterial is not considered environmentally dangerous when compared with the AF biocides, but our results indicate the importance of knowing its potential toxicity against a wide set of organisms in order to improve the manufacturing process of the engineered nanomaterial, as recently proposed by Kaczerewska et al. [34].

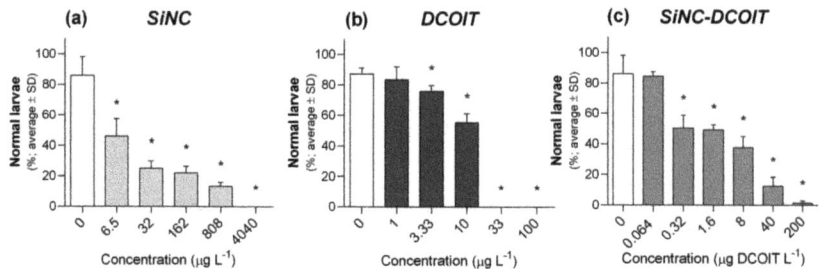

Figure 2. Larval development rates observed in *P. perna* embryos exposed to (**a**) SiNC, (**b**) DCOIT, and (**c**) SiNC-DCOIT. Asterisks (*) indicate significant differences relative to the control (p < 0.05). Data are presented as average ± standard deviation (SD).

Regarding the tested biocide, the development of mussel embryos was significantly affected at 3.3 μg L^{-1} of DCOIT (LOEC). Above 33 μg L^{-1} (Figure 2b) no tested organisms reached a well-developed veliger larvae stage. The embryotoxicity of DCOIT to *P. perna* is in agreement with previous findings for other bivalve species from temperate regions, namely, the mussel *Mytilus edulis*, for which Bellas et al. [35] and the USEPA Office of Pesticides Programs [36] reported 48 h-EC_{50} values of 10.7 and 2.7 μg L^{-1}, respectively. Moreover, Shade et al. [10] estimated the DCOIT 48 h-EC_{50} at 6.9 μg L^{-1} for embryos of the oyster *Crassostrea virginica*.

In the SiNC-DCOIT treatment, the embryonic development rates differed significantly from the control at 0.32 μg L^{-1} (Figure 2c). Moreover, the estimated 48 h-EC_{50} of SiNC-DCOIT was set on 6.77 μg DCOIT L^{-1} (Table 1), nearly 2-fold more toxic to the embryos of *P. perna* than the free form of DCOIT (12.36 μg L^{-1}). The present findings are the first evidence that SiNC-DCOIT can exhibit a better antifouling efficacy comparing with the non-nanostructured form of DCOIT in the pre-settlement stage of a fouler organism. This can be justified by the very low and slow release of DCOIT from the nanocapsules through diffusion, during the exposure period (undetectable by HPLC with a high detection limit of 240 µg DCOIT L^{-1} [12]). Interestingly, similar findings were previously reported on settlement inhibition assays of other foulers, such as the bryozoan *Bugula neritina*, and the mussels *Mytilus galloprovincialis* and *Brachidontes pharaonis*, in a study comparing the efficacy of other AF nanomaterials containing Zn or Cu pyrithiones (ZnPT and CuPT) and the respective free/dissolved forms [33]. As an example, CuPT immobilized in engineered nanoclays (LDH) can be up to 250-fold more efficient in inhibiting the settlement of the Mediterranean *B. neritina* larvae than CuPT [33]. Naturally, species sensitivity may differ across the worldwide marine ecosystems, and abiotic factors, such as temperature and salinity can affect the performance of AF compounds [33]. Despite more studies are needed with other fouler organisms, it is safe to conclude that the innovative nanomaterials demonstrate a promising antifouling capacity.

3.3. Short-Term Exposure Assay with Juvenile Mussel Stages

The parameters estimated for the studied chemicals on byssus threads production are presented in Table 1. The number of byssus threads secreted by mussels exposed to the different treatments showed no significant differences comparing to organisms from the experimental control, except for SiNC-DCOIT (at 1000 µg DCOIT L^{-1}), in which the number was significantly lower. However, increasing concentrations of DCOIT and SiNC-DCOIT reduce byssus production (Figure 3b,c, respectively). The 72 h-EC$_{50}$ value of DCOIT was estimated at 96.1 µg L^{-1}, being 3-fold more toxic than the SiNC-DCOIT (305.5 µg DCOIT L^{-1}) and one order of magnitude more toxic than the unloaded SiNC (1323 µg L^{-1}). These data confirm the high antifouling efficacy of free DCOIT, which inhibit the attachment of *P. perna*. In temperate and subtropical regions, mussels have been successfully used as a model system to test the antifouling activity of both free-forms and nanostructurated biocides by determining their continued attachment by byssus threads [33].

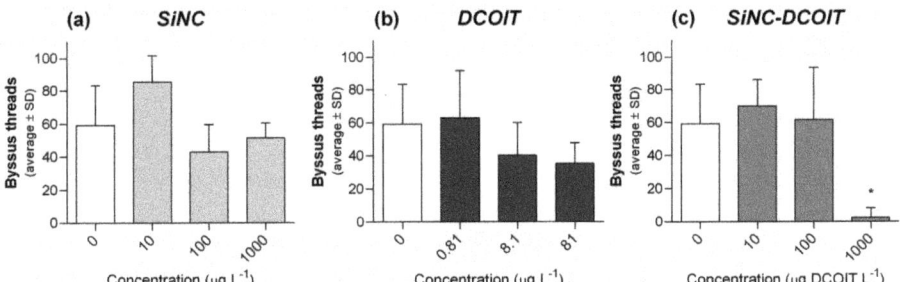

Figure 3. Byssus threads produced by juvenile mussels exposed for 72 h in aqueous solution/dispersion of (**a**) SiNC, (**b**) DCOIT, and (**c**) SiNC-DCOIT. Asterisks (*) indicate significant differences relative to the control ($p < 0.05$). Data are presented as average ± standard deviation (SD).

Regarding to evaluation of the capacity of organisms to survive to aerial exposure after the chemical exposure in water, mussels from the control group survived for approximately 53 h (Table 1). Mussels previously exposed to SiNC treatments survived in air, in average, 40 ± 17 h at 10 µg L^{-1}, 40 ± 21 h at 100 µg L^{-1} and 36 ± 23 h at 1000 µg L^{-1}. The air survival capacity of mussels exposed to unloaded SiNC did not differ significantly concerning the experimental control. SiNC-DCOIT impaired the mussels physiological capacity to survive to a prolonged aerial exposure, set on 33 ± 12 h at 10 µg L^{-1} and 31 ± 11 h at 100 µg DCOIT L^{-1} treatments (not possible to measure on the highest SiNC-DCOIT tested concentration (1000 µg DCOIT L^{-1}) due to the high lethality during the aqueous exposure). Juvenile *P. perna* exposed to dissolved DCOIT presented the lowest survival-in-air rate. In average, the air survival was 31 ± 17, 29 ± 14, and 24 ± 13 h for animals previously exposed to 0.81, 8.1, and 81 µg DCOIT L^{-1}, respectively, being significantly different from the control. It was possible to conclude that the average survival was reduced by less than 50% for mussels exposed to

81 µg DCOIT L^{-1} (Figure 4). Since *P. perna* mussels live mainly at the intertidal level and are periodically exposed to air during tidal cycles, less ability to survive may indicate limited physiological capacity.

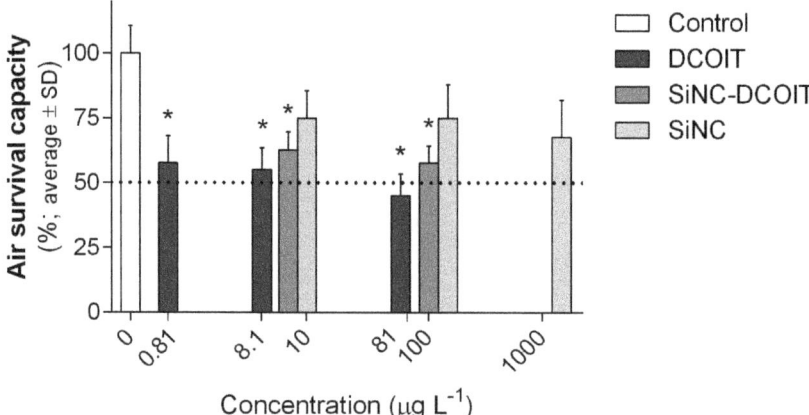

Figure 4. Results of the juvenile mussel survival-in-air after exposition to DCOIT, SiNC-DCOIT and SiNC. Concentration is expressed as DCOIT content (µg DCOIT L^{-1}) in the case of SiNC-DCOIT. Asterisks (*) indicate significant differences relative to the control ($p < 0.05$). Data are presented as average ± standard deviation (SD).

Overall findings indicate that the biocide's encapsulation protected the mussels from significant effects compared with the DCOIT exposure. Similarly, the acute toxicity of DCOIT, in its free form, in the mussel *Mytilus galloprovincialis*, from the temperate region was much higher than the nanostructured form SiNC-DCOIT (72 h-EC$_{50}$ = 1270 µg DCOIT L^{-1} and 38,500 µg DCOIT L^{-1}, respectively; [12]). According to the authors, these mussels close the valves as a defense mechanism to prevent the animals' exposure to contaminants. Thereby, the survival of mussels during the 72 h exposure to free DCOIT solutions can also be related to this mechanism to avoid exposure. In this sense, these individuals possibly remained without oxygenation of gills for a longer time, characterizing physiological effects that reduced their survival capability in a dry environment.

3.4. Environmental Relevance

No information regarding the presence or levels of DCOIT throughout tropical and subtropical zones has been reported in literature. However, DCOIT has been detected in coastal waters of the temperate climate zone in the northern hemisphere, such as Greece [13], Japan [37] or Sweden [38], reaching a maximum of 3.7 µg L^{-1} reported in a Spanish marina [39]. Since this level is above the LOEC value determined for the DCOIT exposure in the fertilization bioassay and above both LOEC values determined in the embryotoxicity test with DCOIT and SiNC-DCOIT, adverse effects on the reproduction of bivalves may occur. In the other hand, the statistical predicted no effect concentration (PNEC) based on L/E/IC 50 values of DCOIT was recently set on 0.2 µg L^{-1} [40], one order of magnitude higher comparatively to the EC$_{50}$ of DCOIT on *P. perna* fertilization. Thus, environmental concentrations of DCOIT may critically impair natural populations of this species, indicating potential ecological risks worldwide. Fonseca et al. also shown that the viability of haemocytes of mussels of *P. perna* are affected by DCOIT exposure as early as 24 h of exposure [41]. It is important to emphasize that the lowest exposure concentration of DCOIT (with environmental relevance) of the present study also caused a reduction on the air survival capacity of *P. perna* mussels by 45%. Since the natural populations of *P. perna* mussels inhabits intertidal rocky shores, a continuous exposure, particularly close to marinas or harbors, may cause incapacity of mussels to properly cope with the presence of DCOIT. Thus, a hypothetical decline of the established populations caused by the continuous exposure to DCOIT can have a negative socioeconomic impact in mid and low-income countries. In Brazil, for instance, *P. perna* mussels has both ecological and economical relevance [42], and many natural populations and mussel farms are located close to marinas, being under potential influence of AF compounds, including DCOIT.

In this sense, deleterious levels of DCOIT can foster technological developments to control the biocidal leaching from maritime coatings with environmental and economic benefits for the maritime industry. Recently, Figueiredo et al. [40] demonstrated that the encapsulation of DCOIT was able to promote a 25-fold decrease on the marine hazard of DCOIT in temperate ecosystems shown by the increase of the PNEC values from 0.2 to 5 μg L^{-1} on the conventional DCOIT and SiNC-DCOIT, respectively [40]. In the same direction, the present study demonstrated for the first time the promising antifouling efficacy of the novel nanostructured biocide in early life stages of the mussel *P. perna* together with the reduced toxicity on non-target stages of this neotropical species compared with the conventional DCOIT. These results are explained by the slow and controlled release of biocide in time, as recently demonstrated by Figueiredo et al. [12] in artificial saltwater. Since this is the first ecotoxicological study ever conducted in a tropical species, future research should focus on the holistic assessment of the fate, behavior, toxicity and hazard on the tropical environment of this novel nanoadditive for maritime coatings to avoid the same mistakes of the past when conventional biocides entered in the market without an appropriate assessment of their effects on the environment.

4. Conclusions

This is the first study assessing the ecotoxicological effects of SiNC-DCOIT, a novel AF nanomaterial, in a tropical marine species. The present findings reinforce the importance of balancing toxicity towards non-target species and efficacy against fouler species when developing a novel AF biocide. Using the mussel *P. perna*, as a model to holistically assess DCOIT in a dissolved and nanostructured forms, it was possible (a) to confirm that DCOIT is very efficient inhibiting the attachment of juvenile *P. perna*, nevertheless extremely toxic towards gametes and very early stages of the tested mussels; (b) to demonstrate that SiNC-DCOIT can be even more efficient than DCOIT, by preventing the formation of the *P. perna* veliger larvae, which can be the critical to control the undesired settlement of such fouler species on coated immersed surfaces; (c) to corroborate the lower toxicity of SiNC-DCOIT (comparing with DCOIT), now on fertilized gametes and juveniles of *P. perna*, thus ensuring the ecological success of the natural populations of this species.

Therefore, SiNC-DCOIT can be regarded as promising AF additive for maritime coatings that poses lower environmental risk while can successfully tackle the adhesion of larvae of *P. perna* thanks to the controlled biocidal release of this engineered nanomaterial. Future integrative studies would unveil the holistic effects on tropical marine biota and environment.

Author Contributions: Conceptualization, R.M. and F.C.P.; validation, R.M., D.M.d.S.A. and F.C.P.; formal analysis, J.V.N.d.S. and F.C.P.; investigation, J.V.N.d.S., R.M., M.K.F., B.G.d.C., M.B.M.d.P.e.S. and F.C.P.; resources, R.M. and F.M.; writing—original draft preparation, J.V.N.d.S. and F.C.P.; writing—review and editing, J.V.N.d.S., R.M., M.K.F., B.G.d.C., M.B.M.d.P.e.S., D.M.d.S.A. and F.C.P.; supervision, R.M. and F.C.P.; project administration, R.M. and D.M.d.S.A.; funding acquisition, R.M. and D.M.d.S.A. All authors have read and agreed to the published version of the manuscript.

Funding: This project was carried out in the framework of the bilateral project "Exposure and bioaccumulation assessment of anti-fouling nanomaterials in marine organisms from temperate and tropical waters" funded by Fundação para a Ciência e a Tecnologia—FCT (ref. 4265 DRI/FCT), Coordenação de Aperfeiçoamento de Pessoal de Nível Superior—CAPES (Process #88881.156405/2017-01) and the project NANOGREEN (CIRCNA/BRB/0291/2019) funded by national funds (OE), through FCT, I.P. RM is funded by national funds (OE), through FCT, in the scope of the framework contract foreseen in the numbers 4, 5 and 6 of the article 23, of the Decree-Law 57/2016, of August 29, changed by Law 57/2017, of July 19 (CEECIND/01329/2017). JVNS (grant CNPq #46593), MBMPS (grant CNPq #53805) and DMSA (grants #311609/2014-7 and #308533/2018-6) is funded by National Council for Scientific and Technological Development—CNPq. MKF (grant FAPESP #2016/24033-3), BGC (grants FAPESP #2017/10211-0 and #2019/19898-3) and MBMPS (grant FAPESP #2018/25379-6) is funded by São Paulo Research Foundation—FAPESP. FCP is funded by CAPES (grant CAPES #88887.124100/2016-00). Thanks are also due to the financial support to CESAM (UIDP/50017/2020 + UIDB/50017/2020), through national funds.

Acknowledgments: We acknowledge Smallmatek by providing all test materials.

Conflicts of Interest: The authors declare no conflict of interest. The funders had no role in the design of the study; in the collection, analyses, or interpretation of data; in the writing of the manuscript, or in the decision to publish the results.

References

1. Chen, L.; Lam, J.C.W. SeaNine 211 as antifouling biocide: A coastal pollutant of emerging concern. *J. Environ. Sci.* **2017**, *61*, 68–79. [CrossRef] [PubMed]
2. Evans, S.M.; Birchenough, A.C.; Brancato, M.S. The TBT Ban: Out of the Frying Pan into the Fire? *Mar. Pollut. Bull.* **2000**, *40*, 204–211. [CrossRef]
3. Fernandes, J.A.; Santos, L.; Vance, T.; Fileman, T.; Smith, D.; Bishop, J.D.D.; Viard, F.; Queirós, A.M.; Merino, G.; Buisman, E.; et al. Costs and benefits to European shipping of ballast-water and hull-fouling treatment: Impacts of native and non-indigenous species. *Mar. Policy* **2016**, *64*, 148–155. [CrossRef]
4. Voulvoulis, N. Antifouling Paint Booster Biocides: Occurrence and Partitioning in Water and Sediments. In *Handbook of Environmental Chemistry: Water Pollution*; Konstantinou, I.K., Ed.; Springer: Berlin, Germany, 2006; Volume 50, pp. 155–170.
5. Arai, T.; Harino, H. Contamination by Organotin Compounds in Asia. In *Ecotoxicology of Antifouling Biocides*; Arai, T., Harino, H., Ohji, M., Langston, W.J., Eds.; Springer: Tokyo, Japan, 2009; pp. 61–74.
6. Perina, F.C.; Abessa, D.M.S.; Pinho, G.L.L.; Fillmann, G. Comparative toxicity of antifouling compounds on the development of sea urchin. *Ecotoxicology* **2011**, *20*, 1870–1880. [CrossRef]
7. Castro, Í.B.; Westphal, E.; Fillmann, G. Third generation antifouling paints: New biocides in the aquatic environment. *Quim. Nova* **2011**, *34*, 1021–1031. [CrossRef]
8. [USEPA] United States Environmental Protection Agency. Presidential Green Chemistry Challenge: 1996 Designing Greener Chemicals Award Rohm and Haas Company (now a subsidiary of The Dow Chemical Company). Designing an Environmentally Safe Marine Antifoulant. Available online: https://www.epa.gov/greenchemistry/presidential-green-chemistry-challenge-1996-designing-greener-chemicals-award (accessed on 27 October 2020).
9. Thomas, K.V. The environmental fate and behaviour of antifouling paint booster biocides: A review. *Biofouling* **2001**, *17*, 73–86. [CrossRef]
10. Shade, W.D.; Hurt, S.S.; Jacobson, A.H.; Reinert, K.H. Ecological Risk Assessment of a Novel Marine Antifoulant. In *Environmental Toxicology and Risk Assessment: Second Volume*; STP1216-EB; Gorsuch, J.W., Dwyer, F.J., Ingersoll, C.G., LaPoint, T.W., Eds.; ASTM International: Philadelphia, PA, USA, 1993; Volume 2, pp. 408–1993.
11. Chen, L.; Xu, Y.; Wang, W.; Qian, P.Y. Degradation kinetics of a potent antifouling agent, butenolide, under various environmental conditions. *Chemosphere* **2015**, *119*, 1075–1083. [CrossRef]
12. Figueiredo, J.; Oliveira, T.; Ferreira, V.; Sushkova, A.; Silva, S.; Carneiro, D.; Cardoso, D.N.; Gonçalves, S.F.; Maia, F.; Rocha, C.; et al. Toxicity of innovative anti-fouling nano-based solutions to marine species. *Environ. Sci. Nano* **2019**, *6*, 1418–1429. [CrossRef]
13. Sakkas, V.A.; Konstantinou, I.K.; Lambropoulou, D.A.; Albanis, T.A. Survey for the occurrence of antifouling paint booster biocides in the aquatic environment of Greece. *Environ. Sci. Pollut. Res.* **2002**, *9*, 327–332. [CrossRef]
14. Tsunemasa, N.; Yamazaki, H. Concentration of antifouling biocides and metals in sediment core samples in the northern part of Hiroshima bay. *Int. J. Mol. Sci.* **2014**, *15*, 9991–10004. [CrossRef] [PubMed]
15. Martínez, K.; Barceló, D. Determination of antifouling pesticides and their degradation products in marine sediments by means of ultrasonic extraction and HPLC-APCI-MS. *Anal. Bioanal. Chem.* **2001**, *370*, 940–945. [CrossRef] [PubMed]
16. Jung, S.M.; Bae, J.S.; Kang, S.G.; Son, J.S.; Jeon, J.H.; Lee, H.J.; Jeon, J.Y.; Sidharthan, M.; Ryu, S.H.; Shin, H.W. Acute toxicity of organic antifouling biocides to phytoplankton *Nitzschia pungens* and zooplankton *Artemia* larvae. *Mar. Pollut. Bull.* **2017**, *124*, 811–818. [CrossRef] [PubMed]
17. European Chemicals Agency. 4,5-dichloro-2-octyl-2H-isothiazol-3-one. Available online: https://echa.europa.eu/substance-information/-/substanceinfo/100.058.930 (accessed on 27 October 2020).

18. Geiger, T.; Delavy, P.; Hany, R.; Schleuniger, J.; Zinn, M. Encapsulated Zosteric Acid Embedded in Poly [3-hydroxyalkanoate] Coatings—Protection against Biofouling. *Polym. Bull.* **2004**, *52*, 65–72. [CrossRef]
19. Hart, R.L.; Virgallito, D.R.; Work, D.E. Microencapsulation of Biocides and Antifoulingagents. U.S. Patent 7,938,897, 10 May 2011.
20. Szabó, T.; Molnár-Nagy, L.; Bognár, J.; Nyikos, L.; Telegdi, J. Self-healing microcapsules and slow release microspheres in paints. *Prog. Org. Coat.* **2011**, *72*, 52–57. [CrossRef]
21. Zheng, Z.; Huang, X.; Schenderlein, M.; Borisova, D.; Cao, R.; Möhwald, H.; Shchukin, D. Self-healing and antifouling multifunctional coatings based on pH and sulfide ion sensitive nanocontainers. *Adv. Funct. Mater.* **2013**, *23*, 3307–3314. [CrossRef]
22. Avelelas, F.; Martins, R.; Oliveira, T.; Maia, F.; Malheiro, E.; Soares, A.M.V.M.; Loureiro, S.; Tedim, J. Efficacy and Ecotoxicity of Novel Anti-Fouling Nanomaterials in Target and Non-Target Marine Species. *Mar. Biotechnol.* **2017**, *19*, 164–174. [CrossRef] [PubMed]
23. Maia, F.; Silva, A.P.; Fernandes, S.; Cunha, A.; Almeida, A.; Tedim, J.; Zheludkevich, M.L.; Ferreira, M.G.S. Incorporation of biocides in nanocapsules for protective coatings used in maritime applications. *Chem. Eng. J.* **2015**, *270*, 150–157. [CrossRef]
24. Maia, F.; Tedim, J.; Lisenkov, A.D.; Salak, A.N.; Zheludkevich, M.L.; Ferreira, M.G.S. Silica nanocontainers for active corrosion protection. *Nanoscale* **2012**, *4*, 1287–1298. [CrossRef]
25. Lourenço, C.R.; Nicastro, K.R.; McQuaid, C.D.; Chefaoui, R.M.; Assis, J.; Taleb, M.Z.; Zardi, G.I. Evidence for rangewide panmixia despite multiple barriers to dispersal in a marine mussel. *Sci. Rep.* **2017**, *7*, 10279. [CrossRef]
26. Pierri, B.S.; Fossari, T.D.; Magalhães, A.M.R. The brown mussel *Perna perna* in Brazil: Native or exotic? *Arq. Bras. Med. Veterinária Zootec.* **2016**, *68*, 404–414. [CrossRef]
27. Zaroni, L.P.; Abessa, D.M.S.; Lotufo, G.R.; Sousa, E.C.P.M.; Pinto, Y.A. Toxicity Testing with Embryos of Marine Mussels: Protocol Standardization for *Perna perna* (Linnaeus, 1758). *Bull. Environ. Contam. Toxicol.* **2005**, *74*, 793–800. [CrossRef] [PubMed]
28. United States Environmental Protection Agency. *Short-Term Methods for Estimating the Chronic Toxicity of Effluents and Receiving Waters to Marine and Estuarine Organisms*, EPA-821-R-02-014, 3rd ed.; United States Environmental Protection Agency: Washington, DC, USA, 2002.
29. Associação Brasileira de Normas Técnicas. *Aquatic Ecotoxicology—Short Term Method Test with Bivalve Embryos (Mollusca—Bivalvae)*; NBR 16456; Associação Brasileira de Normas Técnicas: Rio de Janeiro, Brazil, 2016.
30. Bellas, J. Toxicity of the booster biocide Sea-Nine to the early developmental stages of the sea urchin *Paracentrotus lividus*. *Aquat. Toxicol.* **2007**, *83*, 52–61. [CrossRef]
31. Kaczerewska, O.; Martins, R.; Figueiredo, J.; Loureiro, S.; Tedim, J. Environmental behaviour and ecotoxicity of cationic surfactants towards marine organisms. *J. Hazard. Mater.* **2020**, *392*, 122299. [CrossRef] [PubMed]
32. Gutner-Hoch, E.; Martins, R.; Maia, F.; Oliveira, T.; Shpigel, M.; Weis, M.; Tedim, J.; Benayahu, Y. Toxicity of engineered micro- and nanomaterials with antifouling properties to the brine shrimp *Artemia salina* and embryonic stages of the sea urchin *Paracentrotus lividus*. *Environ. Pollut.* **2019**, *251*, 530–537. [CrossRef]
33. Gutner-Hoch, E.; Martins, R.; Oliveira, T.; Maia, F.; Soares, A.M.V.M.; Loureiro, S.; Piller, C.; Preiss, I.; Weis, M.; Larroze, S.B.; et al. Antimacrofouling efficacy of innovative inorganic nanomaterials loaded with booster biocides. *J. Mar. Sci. Eng.* **2018**, *6*, 6. [CrossRef]
34. Kaczerewska, O.; Sousa, I.; Martins, R.; Figueiredo, J.; Loureiro, S.; Tedim, J. Gemini surfactant as a template agent for the synthesis of more eco-friendly silica nanocapsules. *Appl. Sci.* **2020**, *10*, 8085. [CrossRef]
35. Bellas, J. Comparative toxicity of alternative antifouling biocides on embryos larvae of marine invertebrates. *Sci. Total Environ.* **2006**, *367*, 573–585. [CrossRef]
36. United States Environmental Protection Agency. *Pesticide Ecotoxicity Database (Formerly: Environmental Effects Database—EEDB)*; ECOREF #344; U.S. Environmental Protection Agency, Environmental Fate and Effects Division: Washington, DC, USA, 2000. Available online: https://cfpub.epa.gov/ecotox/search.cfm (accessed on 11 August 2020).
37. Harino, H.; Kitano, M.; Mori, Y.; Mochida, K.; Kakuno, A.; Arima, S. Degradation of antifouling booster biocides in water. *J. Mar. Biol. Assoc. UK* **2005**, *85*, 33–38. [CrossRef]
38. Readman, J.W. Development, occurrence and regulation of antifouling paint biocides: Historical review and future trends. In *Handbook of Environmental Chemistry: Water Pollution*; Konstantinou, I.K., Ed.; Springer: Berlin, Germany, 2006; Volume 50, pp. 1–15.

39. Martínez, K.; Ferrer, I.; Hernando, M.D.; Fernández-Alba, A.R.; Marcé, R.M.; Borrull, F.; Barceló, D. Occurrence of antifouling biocides in the spanish mediterranean marine environment. *Environ. Technol.* **2001**, *22*, 543–552. [CrossRef]
40. Figueiredo, J.; Loureiro, S.; Martins, R. Hazard of novel anti-fouling nanomaterials and biocides DCOIT and silver to marine organisms. *Environ. Sci. Nano* **2020**, *7*, 1670–1680. [CrossRef]
41. Fonseca, V.B.; Guerreiro, A.S.; Vargas, M.A.; Sandrini, J.Z. Effects of DCOIT (4,5-dichloro-2-octyl-4-isothiazolin-3-one) to the haemocytes of mussels *Perna perna*. *Comp. Biochem. Physiol. C Toxicol. Pharmacol.* **2020**, *232*, 108737. [CrossRef] [PubMed]
42. Resgalla, C.; Brasil, E.S.; Laitano, K.S.; Filho, R.W.R. Physioecology of the mussel *Perna perna* (Mytilidae) in Southern Brazil. *Aquaculture* **2007**, *270*, 464–474. [CrossRef]

Publisher's Note: MDPI stays neutral with regard to jurisdictional claims in published maps and institutional affiliations.

© 2020 by the authors. Licensee MDPI, Basel, Switzerland. This article is an open access article distributed under the terms and conditions of the Creative Commons Attribution (CC BY) license (http://creativecommons.org/licenses/by/4.0/).

MDPI
St. Alban-Anlage 66
4052 Basel
Switzerland
Tel. +41 61 683 77 34
Fax +41 61 302 89 18
www.mdpi.com

Applied Sciences Editorial Office
E-mail: applsci@mdpi.com
www.mdpi.com/journal/applsci

www.ingramcontent.com/pod-product-compliance
Lightning Source LLC
LaVergne TN
LVHW070541100526
838202LV00012B/341